# Communication of Complex Information

## User Goals and Information Needs for Dynamic Web Information

# Communication of Complex Information

## User Goals and Information Needs for Dynamic Web Information

**Michael J. Albers**
*University of Memphis*

2005

LAWRENCE ERLBAUM ASSOCIATES, PUBLISHERS
Mahwah, New Jersey London

Camera ready copy for this book was provided by the author.



Lawrence Erlbaum Associates, Inc., Publishers
10 Industrial Avenue
Mahwah, New Jersey 07430

Cover design by Kathryn Houghtaling Lacey

**Library of Congress Cataloging-in-Publication Data**

Albers, Michael J.
Communication of complex information : user goals and information needs for dynamic Web information.
p. cm.

Includes bibliographical references and index.

ISBN 0-8058-4992-0 (cloth : alk. paper)
ISBN 0-8058-4993-9 (pbk. : alk. paper)
1. Web sites—Design. 2. Communication of technical information. I. Title.
TK5105.888.A374 2004
025.04—dc22 2004050656
CIP

Books published by Lawrence Erlbaum Associates are printed on acid-free paper, and their bindings are chosen for strength and durability.

Printed in the United States of America
10 9 8 7 6 5 4 3 2 1

# Acknowledgements

Many people had an influence on bringing this book about. The original text was a merging of my master's thesis (chair Dr. David Covington) and PhD dissertation (chair Dr. Thomas Barker). Of course, since the word count has almost doubled and content is totally rearranged, I doubt they would recognize it. There are also the anonymous reviewers of the both the first and second version of the proposal I sent off to various publishers. The comments of the first reviewers were instrumental in the complete reorganization of the book to its current form. A great thanks to the editors at Erlbaum, Linda Bathgate for her help throughout the proposal submission process and Nadine Simms for her work as production editor. Thanks also to Dr. Stephen Tabachnick, my department chair, for giving me a course release one semester to focus on writing this book.

I also must acknowledge the Ugly Mug Coffee Studio where I spent too many afternoons drinking too much coffee while writing a large portion of this book. Alas, it closed the week I sent the manuscript to the publisher.

Finally, my thanks to Barbara Mirel for being my initial inspiration in this research area. It was while I was researching my thesis topic that I found her research on addressing open-ended questions which turned me down this strange road of complex situations and defining the user's real questions.

**Michael J. Albers**
April 2004

# Contents

# Preface

## Why I wrote this book

Technical communicators and information designers are increasingly being called upon to address information needs which go beyond providing step-by-step instructions and involve assistance in addressing open-ended situations and problems. Situations and problems that can only be addressed by providing information specific to a situation and presenting it in a way that supports various users' information needs and cognitive processing strategies. Designs that answer these needs require understanding the complexity of the entire contextual situation and providing answers to users real-world questions. The design must assist users in resolving ambiguities inherent in the incomplete information found in complex real-world situations.

## The intended audience

This books helps develop a foundation for analysis and design which supports the approaches to providing the complex information which people require to address real-world situations. A foundation which addresses the issues through a lens of primary audiences of this book: technical communicators, human-computer interaction designers, and information designers. However, I tried not to ignore the needs of the secondary audience of system designers and researchers, specifically researchers in adaptive hypertext and related topics. The work of these people is vital to developing the back ends to the informational systems I envision.

## The issues addressed in the book

The amount of information available for any realistic complex situation overwhelms essentially every user and, for that matter, essentially every designer tasked with presenting the information. In 2000, Outsell projected the market for creating information-content at over $140 billion per year. That figure refers to generating and maintaining the information itself, and not the typical knowledge management estimates that include software and information technology costs. Interestingly, how we can go about providing that information in a coherent and usable format remains very much an unsolved problem. The design issues revolve around how to effectively choose, structure, format and display the information to allow easy access.

Building an effective design that addresses complex information needs requires understanding requires looking at research from the fields of psychology, sociology, human computer interaction, and technical communication and developing a complete picture of the situation. Constantine and Lockwood (1999) describe this idea by stating:

> User interface architecture refers to the overall structure, the total organization of the user interface, not merely its detailed appearance. User interface architecture is not about the individual buttons and drop-down lists or eve the sundry screens and dialogues on which they appear, but rather about how all these things are integrated into a complete system that makes sense to the user (p. 28)

The ending phrase, "makes sense to the user" drives the entire discussion of this book. The analysis ideas I discuss should not be considered an end in themselves, but rather a means of accomplishing the goal of communicating information that makes sense in the context of the current situation. The value of the analysis arises from the end result of being able to see into users' minds and uncover their wants and needs and how to provide for those wants and needs. Within this book the focus is on the user's goals and information needs, not on a system which, as Allen (1996) criticizes, "much information-system design emphasizes the data contained in the system rather than the users of the system and what they want to do, and that is why there are so many bad information systems" (xxi).

This book takes the view that the content of the information system is the most important thing. As such, it is not a typical human-computer interaction book on web design which often seem to overly emphasize arranging widgets on the screen. Instead, the book considers the analysis which needs to be done before the interface is designed and before content is created. It strives to give a way of clearly understanding how the user thinks and what the user needs so that interface operation, content, and presentation can maximize their respective potentials in communicating with a user.

The book begins the development of a theory for information systems that present information in an online environment (in general, I assume web-based systems) when the overall issues and user goals are highly complex. These are systems that have a primary focus of providing information, as opposed to e-commerce web sites designed to sell products, or data-entry sites (or web applications) designed to collect information. Applicable examples would be sites such medical information sites or intranet-based business reports; basically any site that has lots of complex information relevant to a situation which a person needs to understand.

Most of current thinking on how to develop systems which I just described tend to fall into two camps: information retrieval and artificial intelligence (both of which will be needed to make the system work) or socio-technical. However, since I believe no technology will ever replace the need to organize and present data, this book does not take either of those approaches, although it does use research from them. Rather, it takes a communication-based approach that is concerned with communicating information to a user and ensuring the user can properly understand what is being communicated. The concern is not with how to retrieve information from the database, but with how to ensure the proper information exists to be retrieved. And to ensure the retrieved information makes sense to the user with respect to the current situation. The model discussed in this book looks at how to decide what information is needed, which, of course, directly maps onto what content the technical communicator be create. As will be discussed in Chapter 1, the computer system is but one part of the five part model (situation, user goals, information needs, people, system); a part that will be treated in many ways as a black box.

Socio-technical system research deals with the analysis and interaction design for highly dynamic, complex systems such a aircraft control and industrial control rooms and provides a major underpinning to the concepts I present in this book. However, this book is not about socio-technical systems per se, the ideas presented here are limited to presentation and display of information via computer screens, with an assumption communicating textual data via an intranet or the Internet. The socio-technical concerns of human motor response and real-time system control do not play a directly relevant part of the discussion here. As a very basic difference, the rapid feedback inherent in systems control simply does not exist in the systems I describe when a user reads information and responds to it in a less time-pressured manner than is acceptable in most socio-technical systems.

On the other hand, socio-technical system theory has great deal to say about understanding how people react within placed in a highly dynamic environment and when faced with open-ended, complex problems. In a complex situation, the problem will almost away include factors or circumstances not foreseen as part of the original analysis. "As a result, information system design cannot be based solely on expected or frequently encountered situations" (Vicente, 1999a, p. 7). The literature on socio-technical systems constantly repeats variations of the preceding sentence. However, much of this research has not crossed the line into web design, software programming, or technical communication, at least for work in these areas which does not directly involve socio-technical system work. However, that work can help provide insight into many questions which plague the effective development of web-based systems. Too many web applications are designed from either the instructional concepts of leading a person to an answer or the data dump idea. (I'll ignore the problem of too much re-inventing of the wheel as new web designers ignorant of previous research findings develop their own way to doing things and end up committing the design atrocities that populate the web.)

Interestingly, the analysis I discuss in this book maybe harder to clearly accomplish than the analysis for a socio-technical system. For example, socio-technical research may discuss the mental models of nuclear power plant operators and how the information displays must match that mental model. But in a corporate setting, the actual system being influenced may be even more complex than a nuclear power plant. Although a power plant has a huge number of interconnected mechanical systems, it still obeys the laws of physics. On the other hand, corporations have people instead of pumps, valves and motors; people respond according to poorly understood psychological principles of which we cannot even begin to capture more than a small part. Unlike people-based research, socio-technical system research can interact with a system by stimulating it: varying the control inputs and seeing what happens (Johannson, 1993). This may happen either by using operations research to design a mathematical model of the system or it may work with the actual system. Either way, the goal is to gain a comprehensive understanding of the system. But once people enter, injecting a stimulus becomes much harder. Ethical and monetary considerations prohibit moving people around or firing a group to see how it effects the remaining people. And if the people know you are monitoring them, that also effects the results.

## *Can we do this now? Maybe.*

Most people view the Internet as a great library, not as a huge shopping mall and see it as a giant information repository. Unfortunately, to continue the library metaphor, the books are scattered in different bookshelves in different buildings all across town. The long term power of communicating information comes not from quickly retrieving books, but in extracting the relevant parts of 20 books and assembling them into a new interactive book customized for the user's current situation and able to change and adjust as the user's needs change. With the advent of XML, web services, adaptive hypertext, and highly efficient information retrieval, we should soon have the technical capability to provide that customized book. It remains an open question how long it will take to develop the information design capability to truly maximize the potential the technology gives us. Unfortunately, compared to effectively meeting the human requirements, the technology requirements seem trivial.

This book describes informational systems that assemble customized information; a communication possibility that has only recently become possible on any practical real-world scale. The required tools, such as XML, are only starting to reach the maturity to actually build these systems. Many case studies and other examples of single sourcing and content management are starting to appear. I have no doubt that within a few years, the required tools will allow these informational systems to be built with ease. A variety of current tools and research already support various components of them; it's the final melding into full scale dynamic systems that has not yet happened. Note I refer to dynamic systems; I'm thinking well beyond simple template-based single sourcing or single sourcing with conditionals for paper or online help. Rather, I'm thinking of dynamic documents where the online help can be customized to fit the user's experience and knowledge level. Most of the ideas I put forth are relevant to current systems and, if applied, will help make them better. Hopefully, this book can help focus some of the research and help lead us to practical user-centered systems that will address the numerous complex situations faced in everyday life. Thus, this book does not attempt to describe how to build them in detail; rather, it lays a foundation of issues which must be considered as we address complex information systems. Most importantly, it works to keep the focus on communicating relevant information and not on the technology.

The production of high quality systems for complex situations lies not the technology, but in the ability of the designer to understand the user's goals and information needs, and create designs which address those needs. The problem of providing quality information content will never be solved by technology. Solving an information problem requires a careful analysis the user goals and information needs, and that analysis to be applied to the design of systems which supply the information in a manner which meets those user goals and needs. It may require extensive technology to support that process, but alone it is insufficient to make it work.

### *What the book does address*

The complex situations addressed in this book are those which involve users interacting with software to obtain information. Whether the software is a web browser or a customized application is irrelevant. The important element is that the person can obtain the required information in the proper format. The goal is not to display a bunch of numbers on a screen but how to display those numbers so they provide information toward achieving a user's goal. *And help in achieving a goal is what the users require from a system—not just a bunch of numbers* (Constantine and Lockwood, 1999).

I realize that this preface may sound like I am advocating dynamic information creation along the lines of Ted Nelson's Xanadu project with everything being connected to everything else. Most definitely not. As nice as Xanadu sounds, I doubt if it could ever have obtained more than an minor level of its original conception. Likewise, I'm hesitant about what I'm reading about the Semantic Web; it seems to focused on letting computers talk to computers, rather than helping people achieve their goals through interacting with computers. What this book works to develop is a clean way of providing answers to focused information topics. It deals with topics and information needs which can be defined before hand as relevant to the user, although the *actual* information content needed or presented will be dynamically changed. It deals with communicating integrated information to users rather than forcing users to build up the information set themselves from multiple web sites or disjoint files within one web site.

This book considers complex situations and the highly dynamic situational context of information, the aspects of the information, and the information interrelationships required in any system which supports a user's wants and needs. It reviews the current research and performs a synthesis of the material from a wide variety of disciplines, primarily drawing from technical communication, human-computer interaction, and psychology. I build a model relating user goals and information needs to system design and provide an argument revolving around the need to understand user goals and information needs within the context of specific situations.

In discussing the model, I lay out a viewpoint and discuss the various dimensions which must be considered when designing the presentation of complex information. I give a framework within which to perform the analysis, but I realize each case is unique. Each case requires the analysis to precede in a slightly different manner and to end up in a slightly different place. My goal is not to provide clear-cut procedures but to provide a solid understanding of what factors influence the analysis in order for designers to adapt them to their specific situation. Hopefully, after reading this book, a content design and development team can develop clear-cut procedures that will fit their specific situation.

There has been no comprehensive treatment on analysis and design which considers the highly dynamic situational context of information, the aspects of the information, and the information interrelationships required to support the fundamental user wants and needs. Cooper in both *About Face* and *The Inmates are Running the Asylum* probably comes the closest. Since complex information is such a large topic, this book cannot hope to be *the* silver bullet of how to communicate information, but I point out current problems and help move us along the road to better, more effective systems. The ideas I discuss in this book do not form a prescriptive method. Anyone who reads and understands my argument should realize the futility of even expecting a prescriptive method. In multiple places I state that there is not a single way of viewing a complex situation and that the information presentation must allow the person to work through it in their own manner. How then could I dictate a single analysis method to give a user flexibility? Instead, this book presents the ideas and concepts that must be considered when designing a system for presenting information in complex situations; however, it does not try to explicitly define the methods required to accomplish the analysis. It does touch upon many methods and gives explanations how they are useful. Although I acknowledge the futility of a single analysis method, I do make an attempt at one in chapter 7, although I must also acknowledge that it is too general for anyone to use if they skip the preceding chapters. But then, I consider this book as laying out a theoretical and philosophical approach to addressing complex situations, rather than a how-to book.

Actually, I see no purpose in trying to layout a new step-by-step approach. In truth, most of the current user-centered methodologies, such as contextual design or scenarios, with only a change in the analyst's viewpoint, are sufficient to handle the situation. User-centered design methodologies are, by their very nature, designed to be applicable to a wide range of situations and different levels of information development efforts. Thus, rather than adding another detailed methodology, I have chosen to provide an additional layer to take into account when performing user-centered design for systems that involve complex situations.

This book strives to be a starting point for designing web sites that address users complex information needs, but many large holes exist in both our understanding of design and our technology which must be filled before we will truly be able to build dynamic systems to support the complex situations I describe.

# Introduction 1

*I would not give a fig for the simplicity this side of complexity, but would give my life for the simplicity on the other side of complexity.*
*—Oliver Wendell Holmes*

*Over the last decade it became increasingly clear that the complexities of solving problems at the intersection of people's lives, technology and business would require a holistic interdisciplinary process that gave equal weight to many skill sets and disciplines. Gone are the days that a single person or discipline could possess the skills and experience necessary to design and deliver useful and usable products and services.*
*—Challis Hodge*

Individuals seek information because they realize that their knowledge about a situation is incomplete and that the information they need can be found within the system. Supporting the user's information needs is the basic requirement of most systems. Thus, most web and paper documents are designed with one of two assumptions:

- They help people who know what they want and who need help accomplishing it, with "it" ranging from buying a new computer to completing a vacation request;
- They help people learn information, such as when a person goes to the web to find details about a medical problem, investing in a company, or learning a new software program.

Designs that address the information needs of the first bullet assume the information requirement is a linear sequence simple tasks. The design can walk users through the process and they go away happy. But this book is not concerned with design for the first bullet; it addresses the problems of supplying information for the second bullet. Sprague (1995) claims that the major value of online information "derives from its ability to expand the scope of information management from facts in the form of data records and databases, to concepts and ideas that are generally captured, stored, and communicated in the form of documents" (p. 33). Implicit within his claim is the value-add of having information that can address complex situations by providing that information to a user at the time it is required.

Here's an adaptive interface for cardiac physicians that describes the type of system I envision being developed by the methods of this book.

> Physicians need to retrieve rapidly the right patient's information, whatever the amount and the type, elementary or composite, the modes of presentation and the semantical diversity of their relationships. For example when reviewing an ECG, the user may want to directly retrieve the past history of drugs prescription without need to go back and browse the totality of the patient's clinical history.
>
> A possible solution is to base the user interfaces on the concept of hypermedia navigation. The latter allows direct access to the relevant data an the adaptation of the data presentation mode and of the screen layout according to the user requirements.
>
> A human/computer interface has to be ergonomic (coherent, concise, reactive, structured and flexible) and customizable to and by the user. It must automatically adjust its look and feel to suit the requirement of individuals or groups of users. User interfaces have to take into account the cognitive characteristics and the psychological behavior of each user. (Ghedira, Maret, Fayn, & Rubel, 2002, p. 219-220.)

In their paper, Ghedira et al. address the problem from a technical approach; a line of research which is making great progress and is essential for implementing usable and functional systems. However, like most adaptive hypertext researchers, they ignore most issues revolving around the content of the system: How should users needs be defined, what are those needs, what information supports those needs, and how should it be presented. Granted, issues about analysis, design, and creation of content are outside the research scope of computer science, while it is squarely at the center of the research scope for information design and technical communication. This book strives to contribute to information design and technical communication by building part of the theoretical foundation needed to develop the content for adaptive systems.

When a person works within a situation that is not as cut and dry as something such as completing a vacation request, their information needs fall in the latter bullet. The users have real-world information needs; they want help directly relevant to solving their needs to address real-world issues that often involve high-level reasoning and open-ended information needs. In other words, it is hard to clearly define exactly what information is needed and when enough has been gathered. The users face a complex situation that require answers to open-ended questions and any linear information presentation sequence breaks down. Forcing a linear sequence onto these situation creates major usability problems with the resultant system. Conklin (2003) referred to these as "wicked problem," contrasting them with tame problems which have straightforward and clearly defined solutions.

If the need for information were a jigsaw puzzle, in a straightforward problem, the person can work with the information in the same way a normal jigsaw puzzle is solved. Each piece is static and never changes. Matching pieces are found and attached. If they happen to fit into a place in the puzzle that is not yet assembled, then two or three connected pieces can be placed aside and inserted later. But in a complex situation, the pieces are not static. Imagine working on a jigsaw puzzle (Figure 1.1) where the pieces laying on the table change over time and fitting two pieces together changes how the other sides of each piece fit into the puzzle. Yet in a complex situation, the user's goals and information needs tend to conform more closely to this latter view than the former static jigsaw puzzle.

Just like the dynamic jigsaw puzzle, real world work is complex and is based on a collection of different systems that interact in complex ways; ways so complex that basically nobody understands the entire system (Beyer & Holtzblatt, 1998). Realistically, the entire sequence can't even be completely defined because of both the difficulties of defining the minute details that make it up and human nature that leads people to jump around within the data. Users come to multiple decision points, and the sequence starts to form a complex web. Allen (1996) considers how people each look at problems differently.

> An individual facing one kind of problem may need to explore a topic area in a different way than an individual facing a different kind of problem. This kind of problem-based flexibility in the presentation of information is not frequently found in current information systems. In part, this inflexibility may be attributed to the data-centered

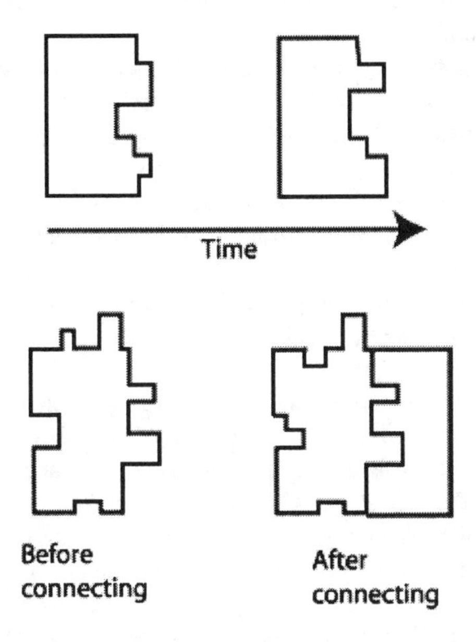


Fig. 1.1. Working a dynamic jigsaw puzzle. The pieces are not static, instead they change both over time and from being fitted together.

> approach, which focuses on the data rather than on the uses to which the data can be applied. It is also true that implementing information systems that can tailor their functioning to the different problem faced by users is far more difficult than implementing systems with less flexibility. (p. 115)

Handling the shift from a linear information model to the ill-structured model requires a design shift. Redish (1990) argues that documentation move up a level and address goals and task repertoires and she recognizes that that emphasis on "higher than discrete task" is crucial for anyone doing everyday work. Accomplishing this requires us to view the creation and presentation of content and its subsequent communication from a humanistic viewpoint, rather than a mechanistic one (Whitburn, 1984). The user goals and information needs must be placed within a proper social and technical contexts and designed to assist people, rather than doing it for them (Belkin, 1980; Ruskin, 2000; Schriver, 1996). The driving force for this shift arises because people, rather than machines, are reading the information. Modern web interfaces frequently deal with the presentation of vast quantities of information, and often address complex situations. To complicate the situation, the basic information must meet diverse needs. Unfortunately, how well the interfaces addresses the user's complex needs, rather providing individual information elements, varies greatly between systems.

Document management or content management systems are one of the current rages in the trade literature. The vendors promise their system can capture all the documents within a company and produce it at will. However, the reality tends to be much more muted. For example, Rein et al. (1997) point out that most of the successful document management applications are relatively simple, well understood work processes (phone directory, library catalogs). On the other hand, large repositories with varied information organization and lack of strict control tend to provide less than hoped for returns. The traditional document management model tries to identify and retrieve a small number of documents out of a much larger collection. With simple situations, a single query usually suffices. In contrast, as the situation becomes increasingly complex, the fundamental goals becomes one of seeking out and learning about information to understand the relationships within clusters of similar documents, and exploit those relationships. Rather than a single query, decision making requires integrating the results of multiple queries (Ebert et al., 1997). The question as shifted from a simple "does this exist" to a much more complex formulations such as "I need to analyze these documents to understand about X. They all discuss X, but which ones contain relevant information? And, more importantly, what is the relevant information for *my* specific needs right now?"

Single sourcing needs some level of content management system running in the background to control the material. Books such as Rockley, Kostur, and Manning's *Managing Enterprise Content: A Unified Content Strategy* (2002) provide high quality information on making use of those systems. This book strives to extend their position to address creating dynamic information systems. Rather than considering how to use a single sourcing systems to create multiple static documents (e.g., print, online help, training), this book looks to move past that view to one of generating documentation customized to each reader's needs. The text presented to a reader conforms to her knowledge level and reading ability so it can best communicate the content.

Web-based information systems are the current best method of creating a useful package of knowledge and delivering it to the people that need the information. Complex situations requiring complex information presentation are a way of life in the modern world. Part of the frustration many people feel searching for information in a computer system arises because the required information they need is hard to integrate into a coherent whole. Terveen, Slefridge, and Long's (1995) work revealed that:

> The pragmatics of knowledge use are critical. Simply recording a factor is not enough; issues such as where in the process knowledge is to be accessed, how to access relevant knowledge from a large information space, and how to allow for change also must be addressed. (p. 3)

Likewise, Mirel's studies have found users have different conceptions of how to accomplish a task. "In actual work settings, users define their own tasks and task needs according to situational demands, not program design" (Mirel, 1992, p. 15). The design of those systems must encompass a total system that revolves around the goals and information needs of a human and supplies information that makes sense within the person's real-world situation. Castel (2002) aptly summed up my argument when he said, "Computing does not merely process information, it commits to a certain representation of information" (p. 30). In this book, I am concerned with designing to improve human performance within specific situations through the interactions between human and computer. In contrast to many existing systems, the complex information systems I envision and describe in this book require the system to efficiently supply information on demand and dynamically adjust that information to maintain the proper detail level.

It is important to understand this book does not consider issues such as natural language generation or other methods of having a computer system generate the information from essentially the ground up. Whereas research continues on those systems, they are a long way from being commercially viable. Instead, the information needs must be meet by creating a proper set of templates and having technical communicators create the content and metadata describing that content.

The examples listed in the sidebar highlight the disconnect between what the person really wants to accomplish and the information the system provides. In each example, the information could exist within the system, but it didn't allow for easy manipulation and integration. Interestingly, when people analyze situations, they enviably must manipulate and integrate information, while they interact with a system that presents it in a limited number of ways (often only one). Cypher (1986) nicely sums up the fundamental problem plaguing current design efforts.

**Complex open-ended information needs**

**Replacing parts on a car**. This one appears deceptively simple. Provide the steps for replacing the part, what more can the user want. Actually, users needs should be addresses more carefully. Ignoring the professional mechanic, consider how a skilled weekend mechanic and a person with basic car knowledge need different levels of presentation and step detail. Consider all these points that can vary:

- **Step detail.** Skilled mechanics need much less detail. Often they can be just generally pointed in the right direction, rather than highly detailed instructions. Yet, other people need to everything spelled out: location of all the bolts, torqueing patterns.
- **Tool usage.** Does the person know how to use a specialized tool or even know that a specialized tool is required? Stopping and trying to find tool instructions can be frustrating.
- **Most procedures address each part separately.** But what if I could enter all the parts I plan to replace even if they are not connected. Then customized instructions could address replacing everything in a single procedure. Something that would be useful if the same piece must be removed to get at both parts.

**Orthodontics information.** During the time I was writing this book, I decided to get braces on my teeth. I went to the web to search for orthodontics information. I have a strong physics and biology background and wanted to see vector diagrams of teeth movement and discussions of bone destruction and formation with respect to the stress caused by the braces. Of course, what I wanted was much more detailed than most people would want or could understand. Different user groups have the same user goal (find out how braces work) but the information needs are very different with respect to detail level and knowledge level. But, in all cases, the question is highly open-ended. There is no single answer to "how do braces work?" instead users need to get information to understand the situation to their satisfaction. (As a side note, I found nothing beyond "do everything your orthodontist says and in two years you'll have a nice smile.")

**Planning a trip to Yellowstone.** A person is planning a trip to Yellowstone. Current travel web sites would give a list of lodges with a reservation link and list of sightseeing sites. Missing is the integration of all this information. Good planning requires knowing the proximity of the selected sites within the park, what can be done at each site, a discussion of traveling between the, how long to allocate for viewing. The person should be able to figure out that two attractions can be done on Monday, but that a third one must be put off to Tuesday because of excessive travel time. Access Travel Guides attempt to address this problem by working with neighborhoods instead of general subject listing as other travel guides do.

> Program designers put a great deal of effort into allowing users to perform single activities well, but considerably less effort goes into allowing users to arrange those activities. (p. 244)

I interpret "allowing users to arrange those activities" to mean providing information useful to address open-ended problems and complex situations which require high-level reasoning. Quality design analysis should not provide only a list of topics describing the data but should also address the real-world situations, goals, and information needs of the users in the appropriate context. Understanding of a complex situation comes from understanding the relationships between multiple pieces of information, then the design must support revealing those relationships. In other worlds, we need better ways to *integrate and communicate the information to a user.*

Interestingly, post hoc analysis of users complaints often finds not that the information isn't there, but that the information can't be found or can't be understood. How many help desk complaints or bug reports get closed with a comment of "documented on page 22" or "working as designed; training issue?" A comparison of two systems which contain the exact same information might find one is considered highly usable and the other is ignored as useless. The poor system might even contain more information but it fails to present it properly to fit the user goals and information needs. There is much more to communicating information than just having the data available. Anyone who makes statements like "the data is right here with some on page 10 and some on 23, what's the problem?" fails to understand just how much information is carried outside the text via the design and the interaction usability. The difference between the good and poor design is not the lack of available information, but the ease with which users can integrate and apply it to the their current situation. Many of the case studies Schriver (1996) discusses results from this same problem of having the information and missing the audience information needs. By missing the audience information needs, effective communication did not occur.

The tools sets are appearing for creating dynamic web content. However, it's very easy to continue to work with a static document model and using those tools to simply add or remove paragraphs. That is not dynamic content. Dynamic information content involves changing the content to reflect either changes in the individual reader's information needs or changes to adjust for different reader's information needs. The fundamental design problem is not helping users find information, it's defining what information the user needs and it must be presented to maximize communication of the content.

The information bottleneck occurs not between the interface and the backend databases (information retrieval research has been very successful at addressing this problem), but between the interface and the user's mind. Although we are getting very good at extracting the appropriate data from the database and moving it from the backend systems to the interface, we need to better understand how to transform the data to information and move that information from the interface to

**Science research articles**

Each week I get a copy of *Science News* in the mail. It contains an assortment of various short pieces of new science research written for the reasonably knowledgeable non-scientist. In other words, unlike the articles that appear in the newspaper, the pieces do not spend most of their time defining terms or discussing how this research is relevant to day-to-day life.

*Science News* is already online at http://www.sciencenews.org/, with the text published as a repeat of the printed version. Consider an online version that follows the concepts I lay out in this book Instead of a 150 word pieces describing a research finding about a gene leading to cancer, the reader could get text that focuses exactly on what they want. For a person with low reading level, the text would be simplified. For a person with an average reading level and basic understanding of genetics, the article would be unchanged from what is written now. For a person with high knowledge level in genetics, but not particularly up on cancer genetic research, the text would provide more technical detail and fill in the holes between general genetics and cancer genetics.

The writer would not write multiple articles, rather the system would have to dynamically assemble the text when the user requested it. Single sourcing and XML are making information content creation for dynamic generation achievable. Information retrieval and adaptive hypertext research are making great strides at figuring out how to get the textual pieces out of a single source database and dynamically manipulated. This book considers about how to assemble those pieces into a coherent article in well-written English (or other language) that provides the information a user needs in an effective format.

The dynamic text I discuss in the previous paragraph is not simple abstracting. During discussions with people about this book, I see people misunderstanding and seeing this concept as little more than what is implemented on many web sites. Namely, provide a 100 word abstract, a 250 word abstract, or the full text. With a dynamic article there is no full text; the text is assembled from many small pieces into a document that fits the current user's goals and information needs.

the user. Figure 1.2 displays the problem of a large stream of information feeding into the interface, but an incoherent, hard to interpret stream emerging. The fundamental problem is that the relevant information coming from the databases is not assembled and organized for a human to mentally process (Hollnagel & Woods, 1983). That assembly and organization of the information for a human is not the job of the information retrieval expert, but an information design expert. The information designer must take the data stream, combine it with an understanding of the user's goals and information needs, and develop an interface that provides the required information in the required format. As Vicente (1999a) points out in the following quote, it's not procedures that people need in complex situations, but flexibility.

> As technological capabilities improve, proceduralized tasks will become increasing automated. As a result, the primary role played by workers in the system is to deal with those situations that cannot be readily anticipate by designers, and thus, automated. Therefore, the need to support worker flexibility and adaptation will only increase in the future. (p. 354)

By concentrating on meeting the users' goals and information needs and ensuring the design enhances the awareness of the overall problem situation, the overall effectiveness of the design ensures that the reader can comprehend and act on the information in an efficient manner. Unfortunately, interfaces are inherently difficult to design and more reflective of creative craft than engineering principles (Myers, 1994), although both interface design and documentation continue to move toward engineering-type principles (Weiss, 2002).

This book provides one view of handling the problem of analyzing the situation and presenting dynamic information for complex open-ended situations. The remainder of this opening chapter lays out a model of the situation, user goals, and information needs, and then provides definitions of simple, complicated, and complex situations. As part of that definition, I consider how designing for a complex open-ended situation requires changing from a design objective of defining tasks to a design objective of gaining an in-depth understanding of the user goals and information needs within specific situations. The remainder of this book then expands on those complex open-ended design considerations.

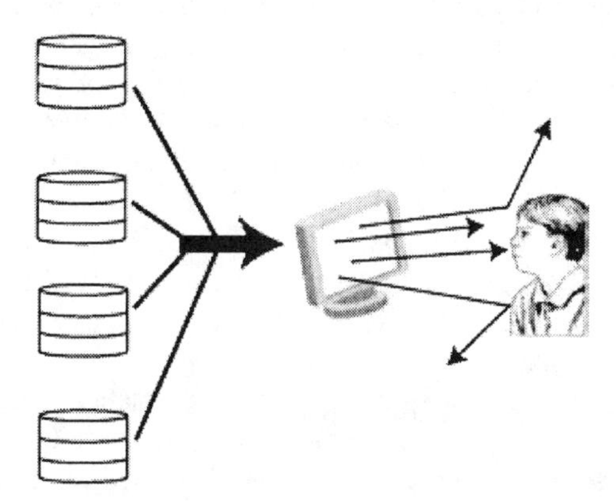

Fig. 1.2. Quality of information flow from the databases to the user. Some information is absorbed and some bounces off (which to the user means it doesn't exist). The less information which bounces off, the better the system communicates with the user.

## Corporate reports

**Example 1**

While I was discussing this book with a potential editor, we were discussing the monthly reports she (as a senior editor) received. Rather than providing any help with interpreting the information in the reports, the report designers took the view of just asking what information was desired and providing a collection of reports that contained it somewhere. A situation which I'm sure most managers can identify with. The issues surrounding how the senior editors and publishers interpreted and worked with the reports each month was deemed too difficult to address and, thus, outside the scope of report design.

**Example 2**

For example, if a marketing analyst for a coffee manufacturer is inquiring into whether a new espresso product is likely to succeed in this specialized market, she needs to view, process, and interact with a wide range of multi-scaled data. To figure out what it will take to break into and become competitive in the high-end espresso market, the analyst will examine as many markets, espresso products, and attributes of products as she deems relevant to her company's goals, and as many as her technical tools and cognitive capacity enable her to analyze. Looking at these products, she will move back and forth in scale between big picture and detailed ("drill-down") views. She will assess how espresso has fared over past and current quarters in different channels of distribution, regions, markets, and stores, and impose on the data her own knowledge of seasonal effects and unexpected market conditions. For different brands and products—including variations in product by attributes such as size, packaging, and flavor—she might analyze 20 factors or more, including dollar sales, volume sales, market share, promotions, percent of households buying, customer demographics and segmentation. She will arrange and rearrange the data to find trends, correlations, and two-and three-way causal relationships; she will filter data, bring back part of them, and compare different views. Each time she will get a different perspective on the lay of the land in the "espresso world." Each path, tangent, and backtracking move will help her to clarify her problem, her goal, and ultimately her strategic and tactical decisions. (Mirel, 2003, pp. 236-237)

By considering report analysis as a complex situation and using dynamic online reports, the report interpretation methods do not have to outside of scope. The information the analysts needs exists. The data used to create the standard report is in a or more relational databases, and textual information such service contracts and memos are in the content management system. The problem is not a lack of data, but a lack of clear methods and techniques to connect that data into an integrated presentation that fits the users goals and information needs.

## Model of situation, goals, and information needs

The fundamental focus of this book is not to explain how to answer simple questions, but how to use web-based systems to address the open-ended questions that arise in complex situations. But before defining simple and complex situations, I want to layout a model that relates the user with a situation, goals, and information needs. As I describe the model, note that the computer system only exists as one of five elements and is relegated to a supporting role, rather than the central role it often occupies in most models. With the decentralization of the computer system, the user goals and information relationships required to understand a situation move to the foreground. As I emphasize throughout the book, these are the most important aspects for the user and, thus, they must be understood when designing to address complex situations.

People constantly find themselves in situations which they want to change in some way. It might be as simple as wanting to get out of the rain to something as complex as having a company that is behind budget and needing to understand the reasons why and develop a recovery plan. Often, rather than resulting in a visible change, the situation change can be purely cognitive, as when a person wants to find out about a particular topic. In this case, the beginning and ending situations are defined by the change in the person's knowledge level about the topic. For example, before you knew nothing about how a database server worked, now you know as much as you want to know.

The common aspect of the examples in the preceding paragraph and the ones presented earlier in the chapter, is that they all involve a person with a goal of obtaining information, interacting with a situation, and changing it. Of course, a person wants to change the situation in such a way that it meets the goal. As part of interacting with the situation, the person acquires data from situation, transforms that data into information, and then, if required, makes a decision about what adjustments to make.

First, here are definitions of the five elements that make up the model shown in figure 1.3. Then I'll discuss how they fit together. Chapters 2 through 6 examine each of these five elements in detail.

Situation — The situation is the current world state which the user needs to understand. An underlying assumption is that the user needs to interact with a system to gain the necessary information in order to understand the situation. In most situations, after understanding the situation, the user will interact with the situation, resulting in a change which must be reflected in an updated system.

Goal — The real-world change of the current situation or understanding of the current situation that the user is trying to achieve. Goals can consist of sub-goals, which are solved in a recursive manner. Goals should be considered from the user-situation viewpoint (what is happening and what does it mean to the user), rather than the system viewpoint (how can the system display a value for x). The system provides a pathway for the user to obtain the information to achieve the goal.

Information need

The information required by the user to achieve a goal. These information needs often require both information to initially understand the situation and to access that the goal was achieved. A major aspect of good design is ensuring that the information is provided in an integrated format that matches the information needs required by the user goals. Although a perfect system always contains the information needed to meet a goal, in reality, the design must compensate for incomplete information.

People

The central person or persons are the ones with the intention of interacting with the situation via the information system. The information system provides them with information relevant to their goals, which they use to understand the situation.

Other people which may be involved in the situation, but not working directly interacting with the information system should be considered as part of the situation itself. In a non-cooperative or hostile situation, other people may be working to achieve their own goals that are in opposition to the user's goals.

System

In general, the system is defined as a computer (web) based method of providing a user with information about the current situation. The common use of system can mean either just the software system, including associated databases, hardware, and so forth, or a combined view that takes into account the software, the hardware, and the user. This book considers the system as the software and hardware.

Too many systems seem to have a focus that assumes that the problem has been defined and all that is needed are the details to address the problem (Allen, 1996). This assumption holds true for simple situations but fails for complex ones. An objective of this book is to explain how to tease out the goals and information needs of ill-defined problems.

If the model could show a time element, the situation would be shown changing based on both situation development and user interactions, with the user's actions aimed at accomplishing the goals representing the desired transformations (Figure 1.4). Of course, in most real-world situations, achieving the overall goal requires a recursive process of having the situation undergo numerous smaller transformations.

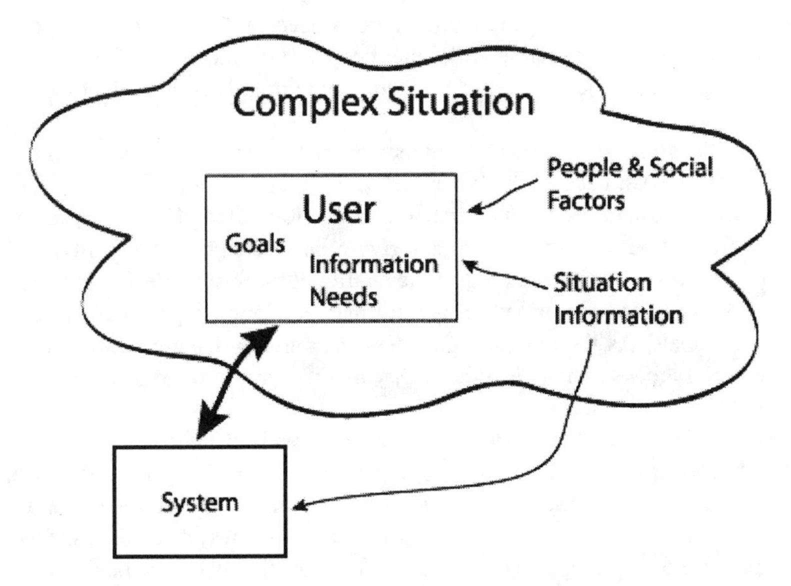


Fig. 1.3. Model of situation, people, goals, and information needs. Note how the information from the system forms a vital component of how the user understands the situation, but the system itself is outside the situation. However, the user sits squarely in the middle of the situation.

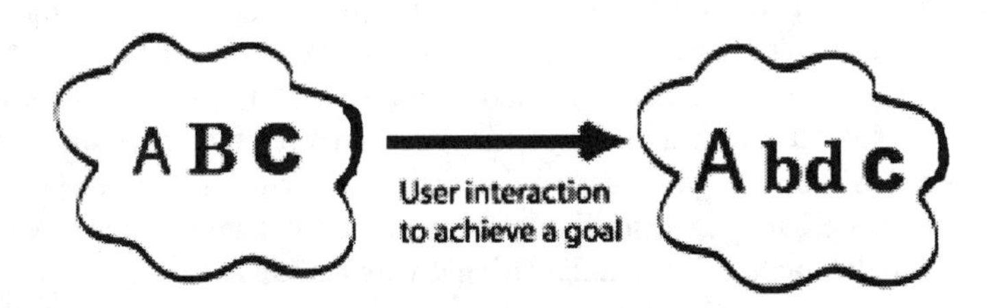


Fig. 1.4. Changes in situation factors over time. Based on both the user interactions to achieve a goal and simply the passage of time, individual factors within the situation change. The user needs to be able to comprehend the initial state, transition states, and final state of each factor to ensure the change achieved the goal.

In this situation-centered model, the emphasis is on getting information to the user so they understand the situation and can achieve the goal (i.e., transform the situation from the beginning state to the ending state). Any situation contains an overabundance of data, mostly irrelevant to the current user goals, that complicates the task of extracting data. The person needs to sift through the available data, decide what is relevant, and use it to achieve the goal. However, the system can only provide part of the information the person needs. Other information comes from sources such as interacting with other people in the situation. There are also confounding aspects such as political or economic limitations that affect valid interpretations of the information.

The system receives data from the situation and feeds it to the person, but does not affect the situation. The system takes in data both internal to the situation and external to it and provides the person with information. The irrelevant data must be filtered out and important data give proper salience. The ability of the person to use that information reflects the quality of the system design to provide integrated and coherent information and information relationships directly relevant to the user goals. As the situation changes, users needs dynamic update of the information. Likewise, they need the ability modify their view to fit their current goals and information needs.

Notice that within this model the system essentially sits outside the situation; while acknowledging this is not true in reality, the model was designed to emphasize the user goals and information needs rather than the more conventional emphasis on system and interaction design. (A true artificial intelligent system could sit within the situation and might replace the person.) The analysis and design must focus on understanding the situation and the information relationships to communication with the user, not on producing a system. Of course, no system can exist in isolation from the situation to which it supplies information, nor can it avoid shaping the person's view of the situation. The interaction design and information presentation exert a strong influence on how a person perceives the information and, consequently, how they come to understand the situation. The exact interaction can be consider in terms of different theories, such as distributed cognitive network, or straturation theory. Regardless of the underlying theory lens used to view the relation of situation and system, the mutual influence is real and unavoidable. An influence which should never be underestimated when making the inevitable development tradeoff decisions. caused by reality conflicting with designer desires. If the system provides information with a bias toward hammers instead of wrenches, the person will come to view the situation with the intention of manipulating it with hammers.

Although this model can work at all levels, from simple requests to highly complex situations, actually analyzing a simple situations in terms of this model is extreme overkill. Discussing how a person uses a mall kiosk to find the coffee store has a before situation of not knowing the store location, and an after situation of the person knowing her location, the store's location, and the route to get there. But defining the situation in terms of this model is not helpful in developing a usable mall kiosk. On the other hand, when the initial situation is a department behind budget, the person needs more information than a simple list numbers of the department line item budget. Here the after situation should be not a new budget plan but a balanced budget. The person needs information to understand the effects of cutting various line items (company policy, other department projects, legal obligations) and how the situation may react to the change (changes in employee moral, actual spending changes). Here the situation exists within a complex set of relationships that must be understood and tracked to ensure the proper goal is achieved. Using a situation-centered analysis helps prevent the designer from falling into a trap of just presenting data on a computer screen and not considering how the dynamic response of the situation to budget decisions will affect the situation and consequentially the data.

## Simple, complicated, and complex situations

Throughout the book, the terms simple and complex will be used to describe different situations. This section provides a brief explanation of what each type of situation means and chapter 2 considers the differences in detail. In the process forming a definition, a third category, complicated situations, which sits between them, will be introduced as an extension to simple situations. The basic distinguishing characteristics are listed in table 1.1. In general, this book focuses on information for complex situations since real-world problems almost always fit the complex category, and not the simple category. Unfortunately, many unusable systems have resulted when the writers, information designers, and system designers try to force fit complex information needs into the simple category.

Of course, situations are never purely simple or purely complex; they form a continuum. Within that continuum the defining lines between simple-complicated and complicated-complex are very hazy. In fact, as discussed in chapter 2, the actual classification is situation and viewpoint dependent. In the final analysis, for many complicated situations, the discussions of this book are highly relevant. It makes no sense for a designer to spend time trying to demarcate whether the current system is complicated or complex; if the concepts of this book apply, then most certainly apply them. A much more usable system should result.

### *Simple, complicated and complex depend on view*

The difference between a simple and a complex system can be the user's view of the system rather than an inherent property of the system.

Most data collections systems aim at appearing simple to the user. An ATM appears simple to the user and only requires the user answer the prompts in a coherent manner. However, to a programmer, the ATM is very complicated with a host of details that must be considered to ensure the accuracy, integrity, and security of the transaction. ATM complexity arises from the various errors which might occur and range from a lack of user funds to the network connection dropping in

Table 1.1. Comparison of simple, complicated, and complex situations.

| | Characteristics | Examples |
|---|---|---|
| Simple | • Single path to a solution exists.<br>• Information requirements to answer a question can be predefined.<br>• Effect on the overall system of a change to one factor can be predicted.<br>• Part of a closed system | ATM (user view)<br>Mall store-finder kiosk |
| Complicated | • Total system is a sum of its parts (a complete description can be given).<br>• Multiple paths to a solution exists.<br>• Information requirements to answer a question can be predefined.<br>• Part of a closed system | Tax software<br>Computers<br>Chess<br>Jet engine repair manual<br>Telecommunications switch documentation<br>Mandelbrot set |
| Complex | • Total system is more than just the sum of the parts (a complete description cannot be given).<br>• Multiple paths to a solution exists.<br>• Information requirements to answer a question cannot be predefined.<br>• Effect on the overall system of a change to one factor cannot be predicted (nonlinear response).<br>• Part of an open system. Difficult to break out and study a section in isolation. | Car buying<br>Report analysis<br>Medical patient record system<br>System design of any realistic complexity<br>Traffic flow on a highway |

the middle of the process. However, even from the programming view, this is still a complicated situation, not a complex one since all the various problems can be plotted out and addressed.

Some design problems are complicated because of their sheer size. For example, the multi-volume reference documents of a major telecommunications switch. The development efforts exists as a complicated situation because of the huge amount of information that must be captured and documented in an organized way. A system to help the technical writers develop the documentation would be highly complex, while a system presenting that information to the end users could be simple. The complicatedness arises from coordinating and tracking the amount of information several writers must handle during development. The end-user information access often consists of closed-ended questions, leading to an end use that can be quite simple: supplying values from a table or a step-by-step maintenance procedure.

On the other hand, the view of simple and complex can reverse when the user needs are pure information. Then the complexity exists within the user view and not the designer/developer view. Actually, as I discuss in a later chapter, the simplicity of the designer view arises from not considering the information relationships, a view that should not arise. A design which takes a simple view fills a database or a web site with information about a topic, for example, cystic fibrosis. This can be rather simple from the content provider's point of view: take information and place it into the site hierarchy. But the person that wants to understand cystic fibrosis because a niece was just diagnosed with it, has a very open-end problem. How much information is there to read? How much should they read? What are the important facts? How do they gain a real understanding and not be totally lost in the medical terminology? What medical knowledge do they possess? What is their reading ability? Are there accessibility issues the web design must consider?

## Complex situations are not a well understood

Various other researchers have considered this topic from various viewpoints. Vicente (1999a) uses the term *situated context,* Suchman (1987) called it *situated action* and Roth, Bennet and Woods (1987) called it *unanticipated variability*. However, compared to studies on simple situations, design for complex situations shows a lack of empirical research. Most of the research has been laboratory-based with simple or contrived problems with limited extensibility to real-world problem solving, a problem repeatedly pointed out by Klein (1999) when he discusses his recognition-primed decision making research which explains how people make a decisions when confronted with a complex situation. The nature of laboratory work focused on trying to resolve various effects forces the experimenter to restrict the situation to single variable manipulation. The problem lies in missing the interactions which occur between multiple variables as they all vary within a situation, interactions that are often too complex to capture in a laboratory setting and, so, are controlled for.

Additionally, our knowledge of the methods for presenting the information to support problem solving is not well researched. That is not to say that there is a lack of opinion on how to define the needs and present the information. However, almost all of what appears in conference proceedings and on various email list groups is opinion or "here's what we did" type of descriptions. Whereas the case studies and best-practice descriptions do form a base on which to build, the lack of rigor prevents an analysis that allows us to figure out why the a particular practice did work and how to make it work in another related situation.

Along with the lack of empirical research to support designing for complex open-ended situations, a comprehensive methodology for analyzing the situation does not exist, instead:

> Much of the activity involved in improving user interface designs has focused on improving methods of entering information. Our interface designs are concerned with making the "input" part of the interface as intuitive as possible for users to provide information to databases and programming algorithms. Yet, equally important to our user community is the information we provide back to them. They view reports and other data that enable them to proceed with actions or make critical decisions before acting. (Hackos, 2002)

Unfortunately, as Vicente (1999a) points out these information situations described by Hackos are almost always open-ended. Once the overall system purpose moves from the well-structured tasks of a closed system to the complexity inherent in an open system, the need for carefully considering the users' goals and information needs and uncovering the relationships within that information becomes paramount, but effectively meeting those goal and information needs remains a major design problem. Lederer and Prasad (1992) pointed out more than ten years ago that a major problem with software projects was overlooked tasks and lack of understanding of user requirements. Unfortunately, articles lamenting these two deficiencies continue to appear (McGovern, 2003). For development of web sites delivering content for complex situations, addressing these issues is paramount for having a usable system. Users want answers, not technology; with dynamic information we finally have a fighting chance of providing that information if the proper upfront analysis occurs.

User-centered methods currently provide the best method to address this design problem. This book does not provide a new methodology, rather it considers how the current user-centered methods can be extended to address complex open-ended situations. It is not so much a problem of needing a new methodology, but of taking the proper viewpoint when using the current ones to accomplish the analysis and design goals.

## Car buying example

In a research study (Albers, 2000), I gave people the situation that their company was going to buy five cars. The subjects were told that their boss wanted them to evaluate and recommend the purchase of a new 4-door car which best met the following.

- Occupant safety
- Vehicle security
- Gas mileage
- Cost of ownership
- Price

The information searching process turned into utter disaster. The subjects were unable to compare/contrast the various features between models and even had trouble finding all the requested information on some of them. They were soon frustrated and making comments about how these web sites were useless and they would just go to talk to the dealers. The basic problem was the subjects found short marketing blubs about each car and had no way of comparing or integrating the information without creating their own tables.

A few months later, I had an interview for a usability and interface design position where I mentioned the study and, consequently, we spent a significant part of the interview talking about designing a hypothetical car-buyer web site. As interviews went, it was a bust (as did the company a few months later in the dot-com meltdown), but later reflection gave me insight into how many people assume a simple situation for design, rather than a complex one.

The entire time it was obvious to me that neither of us were discussing information needs from the same viewpoint. The interviewer wanted me to explain how I would use my complex situation approach to design a system that would recommend a car. Essentially, he expected a wizard and most of his questions revolved around how I would figure out what questions to ask and how to interpret them. I kept explaining that the important part was not a sequence of questions that classified the buyer, but the production of information that helped the user evaluate and decide which features they wanted.

The final result of the system with the simple situation view was to say "the best car for you is a Ford Taurus…click here to order." I also have no doubt that if most people reached a simple statement like "buy a Ford Taurus…click here to order" rather than ordering a Taurus, they would backup and play with their answers until it gave them the car they wanted in the first place. Or the person would express their disagreement with the suggestion because of factors which were never questioned. Perhaps they would make statements such as:

- I don't like that car's body styling.
- Fred had that car and it was nothing but trouble.

- I don't buy Fords.
- (if told to buy a minivan) Yes, I have 3 kids at home, but the twins go to college in a few months, so I don't think a minivan is really required.

The simple situation view starts with a hierarchical tree diagram that the questions prune until only one model remains. The complex situation view does not even attempt to define that hierarchical tree. Instead, it works to uncover what factors a person what to know when they decide between design options (i.e., manual or automatic transmission, leather or cloth seats, towing package), and provides that information so they can make an informed choice.

Car buying can be a complex problem with many factors coming into play and each person balancing those factors differently. Even if two people both want a hot sports car, exactly what they consider critical differs. Rather than an approach that tells someone to buy specific model, use an informational approach that supplies the information needed to help balance out the factors and make an informed decision. The requirements analysis does not define the characteristics of a buyer of different models, but figures out what motivates people to consider/reject different options. The system should provide a description of the pros/cons of 2- versus 4-doors, convertible versus hard top, and how different body styles affect insurance. The goal here is not to say "buy a Ford Taurus...click here to order," but to give 5 models that seem to fit the buyer's needs, describes how they fit those needs, how they differ from those needs, and how the models differ. Most importantly, it must provide easy access to information to help the user refine the judgment about which car to buy.

In short, instead of coming at the design problem with a goal to design a wizard that asks a sequence of questions to answer (a technology solution), design from the idea of what information helps define the required and nice-to-have features (a cognitive/social solution). The design analysis must connect the situation to the user goals and not just state a car model to buy.

| **Simple situation view** | **Complex situation view** |
|---|---|
| **Goal:** tell user what car to buy | **Goal**: provide integrated information to allow a user to make a decision |
| **System action**<br>Ask a sequence of questions:<br>Income?<br>2 door or 4 door?<br>Transmission type?<br>Type of radio?<br>Performance?<br>Hardtop, sunroof, convertible?<br>Display the car that best matches. | **System action**<br>Ask basic demographic information<br>Provide information about features that the user can consider or ignore, as desired. Update the information to fit with the selected options.<br>Display a chart of matching cars that lists the differences and has links back to the feature information. |

## Overview of the remaining chapters

The underlying problem the designer must solve to address how people use complex information lies in defining the information needs required to initially define the goals and understand how the goal and information applies to the situation. But more than just a list of goals and data is required; other requirements include considering the cognitive aspects of information processing, the data interrelationships, and how the data and its interrelationships connect with the user's mental model. Expanding on these concepts and considering how they apply to complex situations make up the remaining chapters of this book.

**Chapter 2: Complex situations**

The chapter begins with an explanation of the three types of situations. The distinguishing characteristics simple, complicated, and complex situations are developed with a view toward clearly laying out a foundation from which issues surrounding complex situation can be developed. The purpose is not to explain how to design for simple or complicated systems, but to distinguish how they differ from complex situations and why different design strategies are required. This chapter then introduces the six characteristics that define a complex situation (no single answer, open ended, multidimensional, historical, dynamic, and non-linear), and provides an extended discussion of each one. The last third of the chapter builds on the complex situation characteristics and how they influence a person trying to understand the situation.

**Chapter 3: User goals**

Chapter 3 examines what makes up the goals in a complex situation and how they relate to information needs. Any person working in a situation has goals which normally define the desired end point. As part of achieving the main goal, multiple sub-goals much be achieved, forming a goal-based hierarchy. By working in terms of goals, this chapter departs from much technical communication literature by stressing user goals and neither task goals or task processes. The focus is on what the person needs to achieve and not defining a specific set of tasks In the process of defining and achieving goals, the person faces constraints from the within the situation constrains the possible goals which can be achieved. The constraints can be cognitive, environmental, or technological, all of which impose different limitations and conditions on user goals. Finally, the chapter examines how goals in complex situations emerge from and evolve with the situation and cannot be predefined, but can be predicted. It wraps up by considering what not be able to predefine goals means to the designers and how the design analysis must proceed to capture the goal space into which the goals will fall. Although the analysis cannot capture the goals, it can capture the types of goals and their general conditions.

**Chapter 4: Information needs**

Chapter 4 extends the discussion began in chapter 3 by examining how people use information to achieve their goals. Goals are the objective but they cannot be achieved without proper information. Achieving a goal is essentially gaining an understanding of and acting on information relevant to the goal and discarding irrelevant information. For complex situations, the information is insufficient; instead, the relationships between various information elements must be understood and presented to the user. Often, understanding a situation derives almost fully from an understanding of the information relationships, rather from simply having the information. As such, analysis and design must determine the relationships and how to best present them to the user.

**Chapter 5: People**

The focus of this book is providing people with information about a complex situation so they can gain an understanding of it. As such, it is imperative to discuss the cognitive factors that influence how people react in complex situations and how those factors relate to achieving goals. Many books that address simple situations ignore or minimize the effect of these cognitive factors. However, as the volume of information and the complexity of the information relationships increase, considerations of people's cognitive processes become paramount for providing effective design. Major factors to consider are the cognitive limitations on how people processes information and how they both form and use mental models of the situation.

**Chapter 6: System**

Whereas the other chapters address what information is required, this chapter addresses how a system might work to present the information to the user. It's main objective is to show how a system should work to provide the user with high-quality information that supports a complex situation. Too many software systems support simple situations and fail to scale for complex situations. This chapter attempts to provide both a list of details to consider for potential implementation and arguments to be used to support the need for designing systems that address complex situations. Like the rest of the book, it focuses on communication of information to people and not a nuts and bolts approach to interface design.

**Chapter 7: Process for addressing complex situations**

This chapter develops an eight step process that can be used to collect the information requirements for complex situations. A reoccurring theme of this book is the situational context of the user goals and information needs and the impossibility of completely defining the entire situation. The process presented in this chapter does not contradict that thesis. Instead, the process formalizes the issues that must be understood at each step and provides an analysis and design framework, but like the rest of the book, it simply cannot present a step-by-step solution. Each step is illustrated with an extended example that runs through the entire chapter.

**Chapter 8: Conclusion**

The chapter summaries the preceding chapters with respect to the model developed in chapter 1. It then looks at the open research questions and near-future technology that must be considered when communicating information.

**Appendix: Creating a visual model**

Development of a visual method of connecting goals with information needs and showing the interrelationships that exist between different goals and different information needs.

# Long example—Astronomical information

This example looks at the complex situation of providing astronomical information to the general reader, loosely defined as a non-professional astronomy. Thus, the audience runs from the novice to highly experience amateur astronomers.

The scenario would be that a large web site is being developed to provide the information. Although this example uses astronomy, it would be rather simple to rewrite most of it to apply to any web site focused on providing the general public with information. For example, medical information, museum exhibits or web sites, environmental information about the benefits of turtles or skunks, or the design of medieval cathedrals or baseball stadiums.

## *Complex situation characteristics*

If providing general readers information about a astronomy is a complex situations, then it must fit the characteristics of a complex situation. This section considers each of the characteristics.

| | |
|---|---|
| No single answer | The information a person is seeking can cover a wide range. Perhaps they just want to understand more about one aspect of the astronomy, such as observing a planet or a galaxy. Or they may want a more detailed explanation about the entire subject. Of course, the entire subject of astronomy is obviously too large to address, but the system should help the person focus their information search. Or maybe they want to know about a particular topic they read about in the newspaper. |
| Open-ended questions | Besides covering a wide range, the questions are posed as open-ended questions. A person wanting a more detailed explanation does not know how much more detailed. Astronomy can run from giving hints on observing current constellation to a highly mathematical description of galaxy evolution. Someplace in the middle a person will decide they have enough information, but that stopping point can't be predefined. |
| Multidimensional approach | How the person views the exhibit depends on their individual circumstances and causes them to approach the information differently. A student interested in collecting information for a school project will want dif- |

ferent information than will a person just getting started in astronomy as a hobby. And both of their approaches will be very different from the experienced amateur, who will have a high, but non-mathematical, knowledge and want details.

Dynamic information — The person's goals and information needs are highly dynamic and are continually refined as new information about a particular topic is revealed. How they understand the various relationships within astronomy change, and understanding which emerges from the interaction between their previous knowledge, current goals, and the information supplied by the system.

History — Why the person is searching for information influences how much they want and how they view it. A person looking to verify something looks at the web page differently from someone learning it the first time or generally refreshing their knowledge. Also, a controversial topic causes people to interpret it different based on their initial views.

Non-linear — As a person learns more about astronomy, they come up with more questions. Finding answers to these questions can lead them down different paths and may also make them go back to previous information and reevaluate it. As they develop a well-formed mental model of the information and its relationships and begin to comprehend the topic, they shift from needing general information to specific points to fill in gaps; a shift which can occur abruptly.

## *Designer and writer issues*

### Scope of astronomical information

Assume the site's task is to provide information about various astronomical events and concepts. While an overall scope statement could be "astronomy information," it's much too broad. That equates to a user saying "Tell me everything about astronomy." What does the person what know about astronomy and how does it need to be explained? The user's goals and information needs are typically much more refined in scope. The scope needs to focus in on the individual goals. It also needs to define aspects such as whether the information should provide comparisons between different astronomical concepts.

### Sample scope statements

- What do astronomer do besides look at the pretty pictures the Hubble Space Telescope sends back?
- What's a gamma ray burst?
- What's a supernova?
- What are giant molecular clouds?

## User groups of astronomical information

There are multiple users groups. As a first cut, this list defines four. A careful analysis could break some of these down further and probably define more.

- Professional astronomers. This group wants highly specialized information and probably wants the raw data, and a mathematical-based presentation. As such, they have totally different goals and information needs than do the other user groups. (This group will be disregarded for the rest of this section.)
- Knowledgeable amateur. A person who enjoys learning about astronomy, but does not require or want an academic presentation. This person probably does not want (and may not be able to follow) any mathematics. On the other hand, they probably know all the basic concepts and not want them to be reiterated.
- Student researcher. A student interested in particular aspects with enough detail to contribute to a research project, but not written at such an academic level so as to be difficult to understand, whether by writing style or assumed prior knowledge. The overall knowledge level is probably less than a knowledgeable amateur, but they may expect a more academic presentation.
- Person looking for general information. This includes people who want a basic understanding but probably have only a sketchy knowledge of concepts. It can also include people looking up information because they saw a TV show and want more information.

Coming out of this analysis would be three distinct groups that would require comparable information: (1) professionals, (2) knowledgeable amateurs and student researchers, and (3) people who want general information. Each of these groups require a substantially different amount and type of information. When it feeds back into the scope definition, it could reveal that 3 different templates would be required to address these groups.

## User goals of astronomical information

The user goals vary between user groups, but with some overlap. To keep this example manageable, only one overarching area will be discussed: giant molecular clouds. The following table lists some of the goals each group might set.

Within each of these goals sub-goals can be defined. The sub-goals exemplify the open-ended nature of complex situations. Consider the goal of understanding the star formation details. Student researcher probably already knows basic star formation, so they only need information that connects generic star formation to the technical details of it affects the molecular cloud. Knowledgeable amateurs may want to go beyond that and have information about how they can observe different aspects with their own telescopes. Or they may want to see how the different telescope styles are assembled with different mirror configurations. The general public normally lacks knowledge of either generic star formation or that it occurs in a molecular cloud.

| **Knowledgeable amateur** | **Student researcher** | **General public** |
|---|---|---|
| How do they form?<br>What parts of them are best for observing?<br>How do they help understand star formation? | How do they form?<br>How do stars form in them?<br>How are they arranged in the galaxy?<br>Size and mass information. | What images can I see?<br>Size information.<br>Why is a Hubble image so different from the radio telescope image? |

## Information needs of astronomical information

All of the user goals raised in the previous section have no complete answer. Users just keep looking at information and setting new sub-goals until they are satisfied. It also shows the multi-dimensional ways people can look at the situation, depending on what they want to know about the specific question.

After defining the user goals, the information relevant to each of them need to be defined. In a hard science, such as astronomy, this can be straightforward if tedious. There does exist a correct answer that satisfies the user goals; however, the answer itself is open-ended. The answer to "What is a molecular cloud?" can run from one sentence to filling a book. In the soft sciences and humanities, even defining a correct answer is harder because various scholars can hold differing and equally valid views. In the end, the designer needs to draw the line someplace before writing the book, but must be sure to not cut it off too short. The user who wants a detailed answer will not be happy with a one paragraph answer, but time and cost prevent too long of an answer.

Factors of user knowledge and reading level also comes into play. Two people can be given the same information, but one can receive it in one sentence and another may need a paragraph. The difference is the writing level and amount of detail required to make the text understandable.

The dynamic nature of the information system allows the content to be varied. For example, when looking for information on telescopes, a general reader would need a description of scope types and magnification. But an experienced amateur knows that information and would consider explanations as distracting noise.

### Information relationships in astronomical information

The relationships convert the text from a mass of data to an interrelated and understandable mass of information. A table of the specifications for a star might make sense to the knowledgeable amateur, but that's because she has deep background knowledge to call on to internally form the relationships. Most people, however, lack that background knowledge and require the system to make them.

Building information relationships is one place were science writing departs from science journalism. In a newspaper article on astronomy, there is almost always a focus on the "why study this" and "how does this affect me, as a person on the street," both of which are not really relationships, but part of the journalistic requirements to build interest and draw the reader in. This information should be within any astronomical system, but it should not be the focus. Most people with a strong interest in astronomy either already know this information or they consider it secondary. Instead, they want to see how this piece of information connects to other pieces of astronomical information.

### Visual model

The appendix contains a partial visual model for this example.

### Presentation

Most astronomy web pages take an encyclopedic approach to presenting information. The reader is presented with a list of topic arranged in a hierarchy. The text under each topic is static and rarely relates to other topics. At best, some of the topics might link to each other.

In a custom dynamic system, the design can go beyond a hierarchical listing of topics with static text. Instead of a collection of short definitions of the parts of a molecular cloud, the information can be presented a hypertext article about the topic. Instead, the text on the molecular cloud can vary based on the reader's knowledge and detail desired. So, two readers with different knowledge levels can receive the same information, but presented with different text that corresponds to their current astronomy knowledge level. This could include one reader receiving different examples or more examples, more explanatory text to clarify terms or phrases, and different terms to replace astronomy jargon with more common terms.

## XML code snippet

This section contains a sample based on XML, but I anticipate that most text markup would be very close to this general model. How to actually write and markup text for a highly dynamic document remain an open question. I especially foresee scaling problems in content development arising from the need to coordinate multiple writers and a large number to text elements.

The text for a topic needs to be written in a manner that supports both including the text into templates and for it to be processed to include the proper content. This example uses a <more> tag to flag the dynamic text elements.

In this example, the paragraph only appears when the knowledge level is medium or low and the detail is high.

```
<more knowledge="medium low" detail="high">
    <para>The dust in a molecular cloud...</para>
</more>
```

Writing separate paragraphs for each combination of knowledge and detail level is obviously not practical to any great extent. Instead, the text inside the paragraph must contain the proper conditional statements to allow for modification. An additional level of markup, not addressed in this example, would replace/expand various terms and phrases with text appropriate for the user's reading ability.

```
<para>The dust in a dense core in a molecular cloud…
    <more knowledge="high medium" detail="high">Pressure waves
        within the dense core...</more>
</para>
<para>Inside the molecular cloud
    <more detail="medium high">pressure waves move through and
```

# Complex Situations 2

*Problems cannot be solved at the same level of awareness that created them.*
*—Albert Einstein*

*There's no solution that applies to every problem, and what may be the best approach in one circumstance may be precisely the worst in another.*
*—G. Wienberg*

This chapter examines what is meant by complex situations and how they can be characterized. The first chapter provided a basic definition that this chapter expands; it focuses on specific issues of complex situations and how they respond to human intervention, by:

- Describing the three types of situations used in this book: simple, complicated, and complex.
- Examining the six characteristics of a complex situation in detail.
- Considering what it means to understand a complex situation and how the six characteristics relate to that understanding.
- Considering how understanding complex situations and its relevant information are context dependent.

## Simple, complicated, and complex situations

The chapter 1 provided a basic definition of the three types of situations used in this book. This section expands that definition and considers the characteristics that define each of these three situations.

### *Simple situations*

An important aspect which separates simple look up of data from operations on complex information is that the answer can be clearly defined and assessed as correct/incorrect. While the answer may vary between situations, for a specific situation the answer is always the same. The situation is closed, meaning it is self-contained and is isolated from outside effects.

Simple situations are generally considered to fit these conditions:

| | |
|---|---|
| A single path to a solution can be predefined | For any particular starting condition, there is one path that leads to a solution. A hierarchical troubleshooting chart attempts to define the single solution by asking questions that prune the tree until only the solution remains.<br>Consider the use of a reference document that contains a table of value for configuring software. Depending on if option A, B, or C is installed, the value may be 3, 4, or 5. But once the installed option is known, a single answer exists and all others are wrong (or at least non-optimal). |
| All possible paths through the situation can be defined | A complete hierarchal tree can be build that describes all possible starting and ending situations. This implies closed-ended questions: any particular question has a single correct answer that can be defined in advance. The answer may vary based on specific conditions (e.g., high/low temperature) within the situation, but for any particular instance, one answer exists. |
| The situation is closed | This is actually a restatement of the first two characteristics. The information requirements and actions to move from one situation condition to another can be predefined because, being a closed system, it is isolated from outside factors. It is irrelevant whether 1 or 40 people enter information, how they feel about entering it, and their working conditions. The system may dutifully record that a value was entered late and, following a pre-programmed rule, assign a penalty. Within its isolated environment, rules can be defined for all conditions and exceptions do not exist. |
| Changes are predictable | A change in one place has predictable effects on the rest of the system. Because all paths through the situation can be defined, the affect of following each path or modifying various conditions can be predicted. This means that all incorrect answers can be analyzed and defined: the effect of entering a 3 instead of a 4 can be predicted and a troubleshooting chart could be built. |

When viewed in terms of complex systems the simple situation can seem almost trivial. For example, the situation at a mall store locator kiosk is "I want to buy running shoes." It is important not to confuse simple situations with being simple-to-implement systems. A personal accounting system might be considered a simple situation since accounting has clearly defined rules, but actually coding an accounting system or teaching someone the rules is decidedly not simple.

Simple situations goals and information needs tend to focus on the skills and activities involved in locating and recognizing a single desired chunk of information, such as checking a stock price, looking up a phone number, or rereading an email received last week.

The information need is for single pieces of data without having to uncover the relationships between data points. Essentially, the user achieves the purpose by going to one place and looking at the text. Marks and Dulaney (1998) have described the cognitive processes involved in a simple look-up, which can be described as the act of visually locating and recognizing the desired information. This information might consist of a word, a group of words, a number, or a graphic image. Most literature on Web search and retrieval tends to focus on simple look-up, as does most technical documentation instruction (i.e., how to set a value of X, performing operation Y).

Analysis methods used to define step-by-step procedures are designed to address simple situations. By design, most task analysis works from the viewpoint and needs of a closed system (Hackos, Hammar, & Elser, 1997; King & Teo, 1996; Rainer & Watson, 1996) with a basic assumption of uncovering a well-defined path the reveals a 1:1 correspondence that between the tasks and system functions. Jobs which consist of well-structured routine actions normally form a closed system; for these jobs, task analysis works well and should be used. For example, a normal data entry system is designed with the assumption that it will operate as a closed system. The system doesn't really care who enters data, can force the data to be entered in the proper format and order, and generates a set of preformatted reports. Benyon counters that this task-based approach is problematic because "a complex relationship exists between data, users, and tasks" (1992, p. 253). What he does not point out, however, is that for most situations this complex relationship is inescapable. If the analyst and designer do not confront it and provide support for it in the interface, then the users will be left to deal with this complex relationship unaided. In short, goal-sensitivity is a fact of life that can be ignored but not escaped.

A wide range of situations do correspond to the simple situation characteristics and should be addresses as such. However, for a wider range of technical documents, their design assumes a simple situation because it is quicker and easier, but is not really appropriate for the user. As discussed in later chapters, these designs result in systems that fail to meet their user's needs.

## Examples of simple situations

**A dictionary**

You only look up one word at a time and expect to find all the information relevant to that word in one place.

**A bank ATM**

The ATM is a common example in many books on task analysis and human-computer interaction (HCI). It provides a good example of a simple situation. It has a small, fixed set of tasks and each can be fully defined. HCI authors like using it both because of its familiarity to readers and the way each task can be defined with only one correct path. The ability to fully define each task and provide one path qualifies it as a simple situation.

**A software reference manual**

Reference manuals are designed to address simple situations. Users are assumed to know exactly what is needed, but doesn't remember the exact value or syntax.

However, if you take a wider view of the reference manual and consider questions about when and why to use a function and what function options are appropriate for various uses, then the situation moves from simple to complex. In the complex situations, most reference manuals fail to provide adequate information.

**Maintenance procedures**

The procedure defines the single correct sequence of steps to accomplish the task. When calibrating a gauge on a control panel. the procedures states to apply a voltage of 3.4 volts across points A and B, and adjust potentiometer 2 until the meter reads 50 units.

### *Complicated situations*

Complicated situations are like simple situations in that they can be fully described, but high levels of interactions and large number of components make us view the situation differently. Cilliers amply sums up a complicated situation:

> If a system—despite the fact that it may consist of a huge number of components—can be given a complete description in terms of its individual constituents, such a system is merely *complicated.* Things like jumbo jets and computers are complicated. (1998, p. viii)

Complicated situations can be generally considered to fit the following conditions:

| | |
|---|---|
| All possible paths to a solution can be defined | A complete hierarchy of paths can be built, although in practice this may be very difficult. The hierarchy typically becomes a web with multiple cross-connections between paths. The more cross-connections within the web, the more the situation moves toward complex. |
| Multiple paths to a solution can be defined | Although a simple situation normally has only one possible correct path from a starting condition to an ending point, a complicated situation can have several. With the increased number of pieces interacting within the situation, the possibility exists for different routes and different methods that accomplish the same goal. However, once a path is chosen, the information needs to accomplish that path can be defined. The existence of multiple paths provides the main distinguishing characteristic between simple and complicated situations. |
| The situation is closed | The information requirements and actions to move from one situational condition to another can be predefined (closed system). Of course, as the complicity increases, clearly defining the limits of the situation can be difficult. As this happens the situation enters the boundary between complicated and complex. |
| Changes are predictable | A change in one place has predictable effects on the rest of the system. Because all paths through the situation can be defined, the affect of following each path or modifying various conditions can be predicted. |

Because the situation can be completely defined, although in reality that might be a extremely complicated task, a complete set of user goals and information needs can also be defined. Systems designed to support complicated situations can still work from the assumption of providing tasks that focus on achieving the predefined goals. Of course, the existence of multiple paths means that completely defining the goals and information needs requires defining all the paths. Unfortunately, designer for a system supporting complicated situations often picks one path and ignores the others; although this might be simpler for a designer it makes the system less useful to a user.

As a situation moves into the hazy area between complicated and complex, the third and fourth characteristics become less applicable as it becomes harder to define the boundaries of the situation. The situation becomes open rather than closed as it becomes influenced by outside factors. It becomes much harder to predict how the system will respond because of the large number of interacting components (a characteristic which is part of the definition of complex situations). In real-world situations, most complicated situations quickly move into the hazy area and can be treated as complex situations. Because of this hazy area, this book generally only discusses simple and complex situations; the applicability to complicated situations can be viewed in terms of whether it is best to treat the situation as simple or complex.

## Examples of complicated situations

**Tax preparation**

Government tax collecting agencies (in the United States, this is the IRS) have created a highly interwoven system of various tax forms used to report different types of income and expenses. Although there may be multiple ways of completing the forms, including choosing to report the same information on different forms, the process can be defined. Tax software is a system that attempts to capture the entire definition.

**Chess**

Chess has small set of rules with the complexity arising ecause the decision tree quickly becomes massive. The problem in not defining the situation or even in how to go about defining it, but in actually generating the path. Because our current computing power prevents completely defining the game solutions, as a situation chess sits well into the hazy area of the complicated/complex boundary.

Chess also serves as a good example of how computers and people work within a situation in very different ways. Computer chess programs, even ones capable of playing at the Grandmaster level, work on a brute force approach of attempting an exhaustive analysis of possible moves. The exact cognitive processes people use to rapidly decide on moves is still a very much open question, but most definitely does not even remotely resemble a computer's brute force approach.

**Jet engine repair**

A jet engine has many components that must all fit together with very close tolerances. However, the entire building and repair process can be full defined and described.

**Telecommunications switch documentation development**

A huge amount of data needs to be tracked and maintained across many writers to produce a large multi-volume set.

### *Complex situations*

Complex situation contains too many factors to be completely analyzed, so it is essentially impossible to provide a complete set of information or fully define the paths through the situation. Complex situations contains lots of ambiguity and subtle nuances of information. That fact, if nothing more, forces the human into the process since computers simply cannot handle ambiguity (although research areas such as fuzzy logic are working on the issue). From the computer's point of view, data is never ambiguous: If it has 256 shades of gray, then it can be assigned to one of 256 little bins. The easiest design method is to not even try to have the system address the ambiguity. The system can display the information and allow the user to manipulate it while *the user* sorts out the ambiguity.

A complex situation differs from a simple one because it is not isolated but is constantly effected by various external effects. The external factors provide a level of unpredictability which greatly complicates an analysis. Cilliers (1998) described a complex situation by saying:

> The interaction among constituents of the system, and the interaction between the system and its environment, are of such a nature that the system as a whole cannot be fully understood simply by analyzing its components. Moreover, these relationships are not fixed, but shift and change. (p. viii)

Complex situations are distinguished by the following characteristics. The next section of this chapter examines each point in greater detail.

No single answer — The required information does not exist in a single spot and a single answer, "the correct one," does not exist. (A person researching a disease can go from one sentence descriptions up to reading medical research journals.) Because there is no single answer, users continues until they acquire "sufficient information" about the situation (Klein, 1999; Simon, 1979), with the definition of sufficient depending on the situation and user goals.

Open-ended questions — An open-end question has a no fixed answer that can be defined in advance and it can be difficult, if not impossible, to even state when the question has been answered completely. The main questions of interest in most situations tend to be open-ended. Whereas systems for simple and complicated situations deal in supporting clearly defined goals and providing information to meet those goals, systems for complex situations work with ill-structured, open-ended information needs which make it difficult to define stopping points for when an information need has been met.

Multidimensional strategies — A person needs to use an evaluation strategy that takes into account multiple factors simultaneously, because the information normally comes from sources and can be looked at in different ways. When engaged in answering open-ended questions in complex situations, a person must evaluate information from multiple sources to address the open-ended questions inherent in the situation. There is no "right way" to achieve the goal or acquire the information. Instead of following a fixed path, users continually adjust their mental paths as new information is presented. As such, initial analysis cannot define a single path, much less fully define the information needs. Both the level of information detail and the appropriate presentation format can change repeatedly while looking at the information, requiring a person to change their evaluation strategy.

Has a history

Complex systems have a history, which implies the system has a future. Something has happened in the past that caused the current information to obtain its current values and can affect the future evolution. The system history determines how the current situation should be presented and what information is salient. Having a history also means that the situation is dynamic; its state and information values are constantly changing and evolving.

## Complicated situation designed as a simple situation

I recently moved from one house in town to another. As part of the move, I called the phone company to get the service switched. The first thing I was told was I would be getting a new phone number. In reality, I wanted two new numbers. Then I asked about DSL availability (at the time neither number had it). Then I asked about setting up the disconnect dates and final billing address on my two previous phone numbers. During this long exchange, I could tell the person was getting lost and frustrated working through all the tasks. When I asked, she remarked that I would not believe how many windows were open on her screen.

Actually, I didn't have much trouble visualizing the system design and understanding where all the windows came from. I'm sure the task analysis said the customer service person needs to: add a new number, check for available options, change address, do a disconnect order, and so on. To make this easy to learn and, supposedly easy to use, each option because a menu option that opened a new window. So the poor person probably had my account information window, two add phone number windows, two check available options windows, two disconnect windows, and two define billing address windows. All on a single monitor that hid most of the windows.

Unfortunately for me, the call center was located in a different city (yes, I asked), because I wanted to observe this system in action.

**Dynamic information** None of the important information has a fixed value, rather it changes continually. Also, the user's goals and information needs are highly dynamic and are continually refined as new information is revealed. The user's view and understanding of the situation emerges from analyzing the situation. External influences can change how the information must be interpreted and the information requirements, thus the analysis must consider what changes might occur and when and why. For users to achieve their goals and information needs, they find the proper level of information, and then relate the information elements to each other to gain a clear understanding of the current situation.

**Nonlinear response** The response of a complex situation to interactions is often nonlinear, meaning it is very sensitive to the initial conditions and influencing factors. Minor changes in starting conditions can result in large differences as the situation evolves and how the changes will affect the situation is often unpredictable. With the dynamic nature of meeting complex information needs and variability of the situation, slightly different starting points can require different information and result in different final results.

Examples of nonlinear response include the sudden slowdown in server response on a computer network or traffic flow on a road. A flock of birds, a school of fish, and the Game of Life all show how systems composed of many elements following simple rules can exhibit a nonlinear response that appears as spontaneous self-organization. The outward actions of none of these groups can be predicted by studies individual elements. No matter how much you know about how a fish swims alone, you still lack the ability to predict the highly organized movement that a school exhibits. A movement, which although it looks like all the fish are communicating, can be modeled with very simple rules. But not rules which can be derived from basic one fish swimming rules.

Although many designs seem to assume otherwise, complex situations tend to be the real-world norm, and thus, provide the incentive for this book. With a susceptibility to external effects, nonlinear response, and a preponderance of ill-structured, open-ended questions, complex situations lose the simple one task-one path relationship; a strict rule-based design becomes impossible. Although a simplified version can have one task-one path, the interactions of both external and internal factors can change the proper task execution and renders the simplified version less than useful.

In a complex situation, a person cannot track all the required information without assistance, thus some type of system is required. A system captures data from the situation, combines it with other data, and presents coherent information that is relevant to the situation and the user's goals. The design assumptions needs to shift from a goal of defining the path and providing the optimal answer to a goal of defining user goals and providing sufficient information to address the user's information needs to achieve the goals. The design analysis must focus on determining and defining the users' informational and psychological needs (Lansdale & Ormerod, 1994) and on how to provide the information "to help people solve problems, rather than directly to solve problems posed to them" (Belkin, 1980, p. 134). In the car buying example discussed at the end of chapter 1, the person interviewing me wanted to directly solve the problem while I wanted provide the necessary information so a someone could address it by themselves. The primary goals of this book is assisting designers and writers to face the task of defining and creating the large amounts of information required for a complex situation when neither the specific questions the users wants answered nor the specific information they wants to see can be specified. (I'll qualify the preceding question by acknowledging that the designer can know in general terms the user's questions and information needs, but not the exact formulation.)

## Examples of complex situations

Complex situations can range from comparing fares for different airlines to reviewing and analyzing corporate reports. In a more involved task, particularly one in which the user wished to solve a problem of some kind—for example, to compare sales figures in different quarters or to troubleshoot a piece of equipment—looking up single pieces of information is only part of accomplishing the task.

### Evaluating corporate reports

To contrast simple and complex situations and the support a system must provide, consider the difference in the view of a data entry clerk and the business analyst. The date entry clerk works within a simple situation: Forms cross his desk and he enters the information into the proper screens of the system. Each value has a certain place to go. Problem forms or incomplete information require outside intervention (redo the form or let the supervisor decide what to do) because it has violated the defined limits of the simple situation. On the other hand, the business analyst who receives the reports from that data works with a complex situation. She is tasked with using that information to gain an understanding of how some aspect of the business if operating and recommending if any adjustments should be done to bring the overall business closer to ideal. The business analyst must bring more information to the analysis than was entered. All the external information which cannot be captured as part of the data entry process, but are required for interpretation may prove more influential than the data. Think about the effects each of the following can have on the analysis of a monthly corporate report: the leading salesman quit, a snowstorm closed the northern stores, a major supplier/purchaser filed for bankruptcy, the company was named in a highly publicized lawsuit, or the numbers are fine but a senior executive is pushing for absurd growth numbers this quarter.

**Medical information**

Consider the question "what is cystic fibrosis?" The most effective answer for the reader can span from one sentence to a huge volume of medical journal articles; the proper answer for a user depends on the situation. A student researching a paper on the topic has very different needs than parents faced with a understanding what their child has been diagnosed with. For the parent, the "what is cystic fibrosis" is not even the real information need. They need information about patient care, disease progression, treatment procedures and options, etc. Most importantly, the explanations must be framed in terms of their medical knowledge level, amount of detail desired, and reading ability. It also must be integrated into a coherent whole, rather than a collection of data snippets. The integration prevents the parent from having to read about disease progression, then read about treatment options, and then have to keep going back to disease progression to understand the treatment options at various stages of the disease.

**Design of a software system of any real-world complexity**

Any designer must be aware of the users and of what information they find useful and not useful. Somehow the provider must know what the reader used, and more importantly, what the reader did not find. Tightly coupled with the idea of what information is used is knowledge of the order of use or want. Beyond the simple process of knowing what was used, the information designer must understand why the information is useful or not useful. This idea has major impact on the interaction between the user and information provider. Only by both understanding what was used and why it was used can the system adequately provide information and evolve in its usefulness rather than simply tossing data over the wall to the reader (Deiberger). Systems that support complex situations require constant design adjustment and modification, as the design problem itself is a complex situation. Failure provide that ongoing design refinement leads down the path of the system failing to meet the user's goals.

## Characteristics of complex situations

The previous section briefly described six characteristics of complex situation. This section examines each of these characteristics in greater detail and considers how they relate to addressing a user's goals and information needs.

### *No single answer*

The open-ended nature of questions posed in complex situations are at odds with giving a concise and complete answer. Because no clear path to move the situation between goal states exists, there is no single answer. Instead, the situation contains an abundance of information both relevant and irrelevant to achieving the user goals. With multiple levels of detail available, users can continue to find information applicable to the question until they enter information overload or it requires a knowledge level they do not possess.

Because there is no single answer, users continue until they acquire "sufficient information" about their understanding of the situation; in Simon's terms they "suffice," (Klein, 1999; Simon, 1979) with the definition of sufficient depending on the situation and user goals. In reaching what is considered sufficient, the user mentally integrates the available information and the information relationships and evaluates how it corresponds to meeting the goals and information needs. The search for sufficient information involves mentally balancing the trade-off being between time to find/analyze information and amount of information available (Orasanu & Connolly, 1993).

Defining what is meant by "sufficient information" becomes difficult for both the designer and the user (Klein, 1999) and depends strongly on the users ability to handle interrelationships between information. Unfortunately, people find it difficult to effectively search for and integrate multiple sources of information, requiring the system to provide the information in a manner which relates to the context of the problem.

Although worded differently, Mirel (1998) follows the same line as Conklin's (2001) wicked problems, when she points out that analyzing complex tasks requires seeing more than a single path.

> This broader view is necessary to capture the following traits of complex tasks: paths of action that are unpredictable, paths that are never completely visible from any one vantage point, and nuance judgments and interpretations that involve multiple factors and that yield many solutions. (p. 14)

Failure to consider Mirel's points often results in ignoring vital elements or under estimating the element's importance. Thus, the first step in designing an effective system should not be to define the data needs, but to define the communication situation in terms of the necessary information processes and mental models of the users (Rasmussen, 1986), as the viewing order and specific information required changes with each situation. Rather than following a fixed path, the users continually adjust their mental path as new information presents itself; the main design task becomes defining and evaluating the user's shifting goals and information needs (Jirotka & Goguen, 1994). Applying these ideas to dynamic information means the best user performance results when the system helps organize thinking rather than suggest or enforce a course of action (Eden, 1988; Yoon & Hammer, 1988b).

The user has goals and information needs focused on obtaining a sufficient mental image of the entire situational context. Part of the sufficient mental image requires seeing the problem from different viewpoints and understanding the interrelationships between the available information. Only when this mental image is adequate can the user make an achieve a coherent understanding and use that understanding in an effective manner. The difficulty in helping form the mental images is because each person works in their own way and requires information presented to fit their current needs (Vicente, 1999a). Dynamic information allows the system to adjust the information to fit the user's needs rather force the user to mentally perform the modification.

As a complication, even in two situations that seem basically identical, workload, time, stress, inexperience at analyzing the situation, presentation order, and other factors affect user's ability to mentally integrate the information (Johnson, Payne, & Bettman, 1988; Klein, 1988; Wright, 1974; Webster & Kruglanski, 1994 ). As a result, users may not consider all relevant information or may not know what information is relevant for the situation (Laplante & Flaxman, 1995). These differences require a designer to understand the various aspects of the situation in enough detail to visualize the various information needs and relationships. Thus, an important aspect of the design analysis must be to define the relationships between the information and how to make the important relationships salient.

Design for complex situations contains an inherent conflict because there is no well-defined answer yet the a design analysis must identify potential answers and predict how/when people will need those answers. No analysis can predict exactly what information a particular user may want or how much information will be viewed, or in what in order it will be viewed before a user feels confident enough to make a decision. But more than just analyzing the approach taken by one user or considering the "best approach," good analysis collects the information needed to support the sub-optimal paths frequently used in real-world situations. Most real-world problems lack a clear overarching answer; instead, people are content to move in the right direction and continue to make follow-up decisions which act as adjustments to the initial decision. Addressing the open-ended questions in complex situation should not be viewed as a problem of providing an answer, but of continuing to supply information to lead the user to a sufficient answer.

## Open-ended questions

An open-ended question has no specific answer, but rather can be examined at multiple levels with more information applicable at each level. The user decides when to stop.

As a user interacts with the situation and gains information applicable to the current open-ended question, the "sufficient information" emerges as a part of that interaction. An acceptable answer varies based on the dynamics of the situation; dynamics that arise because the answer depends on the response of the situation to decisions made in the course of finding an answer. Thus, in answering an open-ended question within a complex situation, the user must rely upon mental reasoning and related knowledge to analyze the information pertinent to the situation. In the end, a user reacts in real-time to address the current situation; the design must provide the means to support this real-time reaction.

With open-ended questions, information only has meaning when viewed with respect to the dynamics of the situation. Unlike a rule-based system which works to answer a closed-ended question, for a open-ended question the essence of the information often exists in the relationships between information elements and not within the individual information elements. For a simple example, if I said I paid $21 at a restaurant last night just for my meal, the information means nothing. Even if I say that it was a good chicken dish, you still can't interpret the value because it lacks a connection to restaurant type. For fast food, the price is absurdly high, yet for a ritzy French restaurant in New York City the price is unbelievably low.

An interesting aspect of analysis for open-ended questions occurs because complete answers cannot be specifically foreseen but the information requirements can be generally anticipated. To complicate providing information, the available information is normally incomplete, so the user must extrapolate from what is available. Hutchins (1995) agrees that people's cognitive activities cannot be completely predefined but that to gain an understanding of them requires:

- Analyzing the methods of accomplishing goals in an open system requires identifying the information needs to cope with unfamiliar and unanticipated events in a flexible manner.
- Identifying the current heuristics and methods used to handle similar problems under the current system. These methods have evolved to fit the current social and environmental constraints, until the reasons for their existence are understood, they cannot be ignored. However, just because they exist under the current way of doing things is no reason to carry them forward. But their reason for existence needs to be identified to make a conscious, informed decision about their relevance to the new system.

### *Multidimensional strategies*

The information in a complex situation must be simultaneously evaluated according to different (often opposing) factors. Consider how in the car buying example, in chapter 1, the study participants needed to base a buying decision on five factors (which form a five dimensional analysis space): occupant safety, vehicle security, gas mileage, cost of ownership, and price. Yet there is no car that maximally satisfies all five at once. For example, a small car may get great gas mileage and have a low cost, but will not offer a high level of safety in an accident.

As a result of people evaluating information with multidimensional strategies, the designer faces more than just handling large amounts of information, but must consider what factors enter into the user's mental analysis. The design must dynamically reflect the multiple dimensions used by a person as they integrate and comprehend the information. A major complication is that a complex situation cannot be fully defined at design-time. Thus, the design for addressing the user goals and information needs requires understanding both the overall requirements, how they vary within changing situations, and the different strategies that different people to understand the information relationships.

An overly simplistic view believes that as long as the information is written down someplace, the system has fulfilled its content obligations. But simply sticking details into a document ignores the multidimensional strategies used to access and process information. Mirel (1998) addresses how users mentally process information through mental schemas when she reviews the results of document design research.

> With our growing awareness of reader's schema [mental model] and multi-dimensional strategies for processing information to create meaning, we can no longer rest confident that so long as information is just given, readers will be informed. (p. 130)

Failure to allow for the multidimensional strategies Mirel mentions often constitutes the problems readers have with finding information. The writer assumed a single strategy would be used for finding or using the data. As a result, when the user approached the problem with a different search method or with a different information need the design hampered finding the required data. Mirel's work contains a strong justification for extensive upfront analysis to understand the users' goals and information needs. The closer the final design comes matching the real-world situational demands, the more useful and accepted the result will be.

## *Has a history*

Complex systems have a history: A sequence of past events has caused the situation to obtain its current information values. Having a history also means that the situation is dynamic; its state and information values are constantly changing and evolving with the past influencing that change. Part of forces changing the shape of the jigsaw puzzle pieces seen in Figure 1.1 come from situation history. More importantly, the history determines the situation's current state and the current state effects its future evolution. The situation history determines how information relationships and the user views the current information As it result, it determines how the current information should be presented so that salient information and its relationships are emphasized. Thus, having a history also implies a future. Because the information is changing, obviously the situation will be different in the future. Fully understanding the situation requires having feel for what those changes will be.

The reason a history is fundamental in complex situations and ignored in simple ones is because the historical events effect the future development of a complex situation but not a simple situation. In a complex situation, the past events help to determine the information relationships and influences how the situation develops. But in simple situations, history is generally irrelevant because situation future changes do not depend on how the situation arrived at its current state. Instead, the current situation can be taken as the starting point for projecting into the future; a projection that is simplified as all paths can be completely defined. Considering the idea of "backing up" or "undo" and how it differs between simple and complex situations. In a simple situation, an action can be undone and the situation returns to the previous state. However, in a complex situation, actions can't be undone because they are not reversible. In other words, simply undoing an action does not ensure the situation returns to its previous condition. For example, think about a poor management decision that results in the employees sending out a flurry of resumes. Even if the management reverses the decision a couple of days later, some good people will leave although they had no intention of changing jobs before the initial decision.

Note that this means bad data can't be ignored or deleted, but must be adjusted for as the situation moves forward. Although transactional systems (data entry system) have the capability back out changes if they are incorrect, people don't operate this way; once they see information, they will always process it at some level even if just to disregard it. Consider how this affects many situations, such as a judge instructing a jury to disregard part of a testimony (Cilliers, 1998, p. 107). Although that part of the testimony is no longer in the official trial transcript, it still exerts some level of influence on the jury. This effect helps to highlight the complex situation inherent in a legal proceeding, because the influence it exerts depends on the circumstances surrounding the disregarded testimony. It is much harder to forget information gained when a witness was crying, or the lawyer was screaming his questions, as opposed to a calm eventless presentation.

Understanding the information relevant to a situation and, more importantly, being able to project future values requires understanding that past. Rapid changes in the past may led to continuing rapid changes. Information which appears quite normal in a single view may be, in actuality, rapidly changing. The time spans to attach to the history depends on the information. Some information routinely varies wildly across short times, but display steady trends. Think of the minute-by-minute or day-by-day display of the stock market versus a week-by-week or month-by-month display. The rapid short term fluctuation can make it hard to distinguish the longer term trends by creating too much noise (Figure 2.1). Of course, even the amount of allowable fluctuation can vary. To a day trader, even a minute-by-minute display of the Dow Jones and S&P values is too high level; a minute-by-minute report on individual stocks is needed. Overly long summations for the history can be also bad. When plotted at five year intervals, the stock market might look to increase steadily, but hidden by the long interval are quarters that were both very high and very low. On the other hand, a five year plot might be fine for a person who has invested for retirement and has no intention of selling.

### *Dynamic information*

Information is dynamic in two different ways: by time interval or presentation. Defining how to properly present information that varies by presentation is an important part of supporting complex situations.

Varies on a short time interval
: The value of the information changes within a time in terval short enough to matter to the situation. For example, production or sales values, or location of resources. This type of information must be captured and updated in the information system in real-time, perhaps by querying a transactional database when a user enters a request.

  As an additional complication, besides the information may change as a result of the interacting with the situation (making decisions). These changes must be tracked and provided to the user with proper feedback.

Varies by presentation
: The information the user sees changes to fit the current user goals and information needs, but the underlying information does not change. For example, the infor-

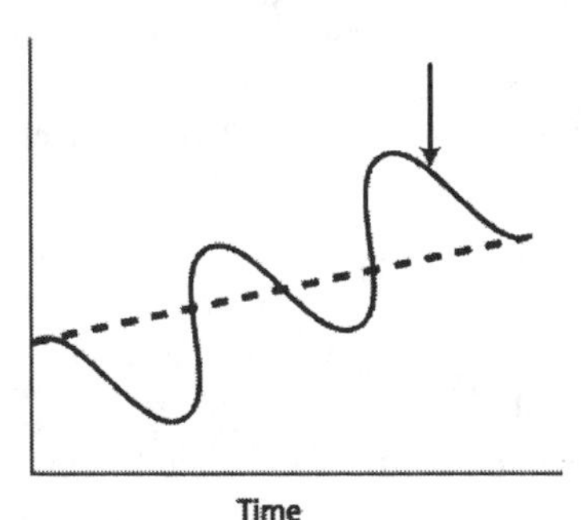


FIG. 2.1. Importance of knowing system history. Both lines have the same average across the graphed time period and same starting/ending points, but a prediction of what how the system will respond in the short term at the time labeled with the arrow would be quite different.

mation presented to two people might be different based on their topic knowledge. Single source technology provides the current best approach to accomplishing this type of dynamic information.

The need for dynamic information arises because the user's goals and information needs continually get adjusted as they interact with the situation. Learning new information can change what user goals and information needs. Thus, what the user needs to understands the situation emerges as a result of interacting with it.

Simple situations tend to have static information or information that changes in a highly predictable pattern. In complex situations, the dynamic information undergoes much less predictable changes, which is discussed in the next section on nonlinear response. Poor design analysis methods try to fix the information or grab a snapshot of the situation. But this ignores the dynamic nature of the very information the analysis needs to collect. Instead of trying to contain, control, or even eliminate changing requirements and user-generated chaos, perhaps we would be better off to prepare, plan, and allow for changing goals and information needs. Accepting the dynamic nature of information means collecting information to handle the following situations, all of which are more realistic for complex situations but require a much more complicated analysis to:

- Handle situations were the information is rapidly changing (dynamic information).
- Handle situations which are highly sensitive to change. How do changes to one data element affect other data elements (interrelationships)? How sensitive are the other elements to change?
- Handle the changes in the situation during the transition period from an action to final result. No complex system can go from one state to a final state instantly. During the transition, many data elements are in flux, the user must not be mislead into thinking these changing elements are new problems which must be addressed. How these change are hard to predict and allow for.
- Handle situations which are pathological or out-of-the-norm. Most people don't turn to informational systems until a major problem or information need has occurred, which by definition means multiple data elements are outside of their normal values.
- Handle error conditions within the situation. Some actions within a situation are incorrect or fail to incorporate all the necessary information. Either way, the situation does not change in a way desirable to reaching the user goals.

### *Nonlinear response*

A complex situation often has a nonlinear response. Nonlinear responses arise in any system that is open, dynamic and constantly evolving, a description that fits essentially every real-world system of interest. Social and environmental interactions are inherently nonlinear, with a structure that is dynamic, changing, and difficult to describe. As a dynamic process, the user goals and information needs in a complex situations is inherently nonlinear.

Mathematicians defines nonlinear systems in very precise terms, using terminology based around equations that are second-order partial differentials. They point out examples such as weather models and movement of air across an airplane wing; situations that can be modeled because the underlying mathematics is understood. However, for complex situations of interest in this book, mathematical equations can't be applied because we simply don't understand the overall psychological problems well enough to define them in mathematical terms. In essence, it comes down to that the system can at times appear highly unpredictable and minor variation in the initial conditions can have major ramifications later on. The impact of seemingly inconsequential nuances of data can have serious effects as they ripple through the situation, causing it to suddenly shift from stable to unstable. The common example is the flap of a butterfly's wings disturbs air currents enough to cause a major storm halfway around the world.

The continual impact of situational influences and the possibility of sudden wild deviations cause problems. Many decision failures occur because of assumptions that the initial conditions were not going to change, or were going to somehow suddenly jump from initial to final state. Complexity arises from high level, nonlinear response of the system to the human interaction.

> People assume and project the future based on linear change or a step function once the new system enters operation, but a real system has a nonlinear change element that might overshoot or undershoot and requires time to settle to a new stable position. (Cilliers, 1998, p. 42)

Besides the psychological issues of people having trouble estimating nonlinear response, a complex system has many external and potential mitigating factors which leads to it being inherently unpredictable over the long term. Each factor introduces a modification into the system which creates a rippling disturbance. Unfortunately, all these disturbances interact with each other producing the nonlinearity of the system. Vicente (1999a) describes the problem as follows:

> First, the very same set of actions will have different effects at different times because the pattern of disturbances will rarely be repeated [note implied reference to system history]. Each situation is a unique context characterized by a different set of contingencies. Second, as a result of this situated context, if the same goal is to be achieved on different occasions, then a different set of actions will be required to achieve the very same task goal. Thus, the action have to vary if the outcome is to remain the same. (p. 76)

Another nonlinear aspect of the complex situations arise during the realistic, gradual change from the initial to final state. The situation can become unstable and swing wildly with unexpected changes. Part of the problem comes from the abrupt changes that factors can have on intermediate states and people's incorrect assumption of a linear response. For example, a group performs an unfavorable, but necessary, task once a month with minor grumbling. But when they are expected to perform the task more frequently, the abrupt shift occurs as the situation goes from minor grumbling to major personnel and productivity problems.

Conklin has examined how different designers addressed working on problems (Figure 2.2). Rather than working in a linear, step-by-step process (exemplified by the waterfall methodology), they constantly shifted from understanding the problem to formulating solutions. However, each designer made the shift from problem to solution at both different times and for different lengths of time. Any person working within a complex situation will display the similar behavior. The problem for the designer is to ensure the system supports it.

Only when the designer accepts the nonlinear nature of complex situations and allows for it in the design does the system have any chance of adequately supporting the information needs of the situation. Designs that allow for control of nonlinear systems must acknowledge the dynamic nature of their information. The information that makes up the initial conditions becomes rapidly outdated; the user must know about the updated situation. The person using the system must be provided enough information to maintain a clear picture of the situation and how it's developing. Each thread of information has its own path and relationships, the design must ensure they can all be tracked. For a business system which is constantly responding to multiple influences, the difficulties of making accurate predictions is obvious. The multitude of factors which simultaneously influence the process lead to multiple threads of information, each basically independent, but weaving around and influencing each other at various ways and at various times.

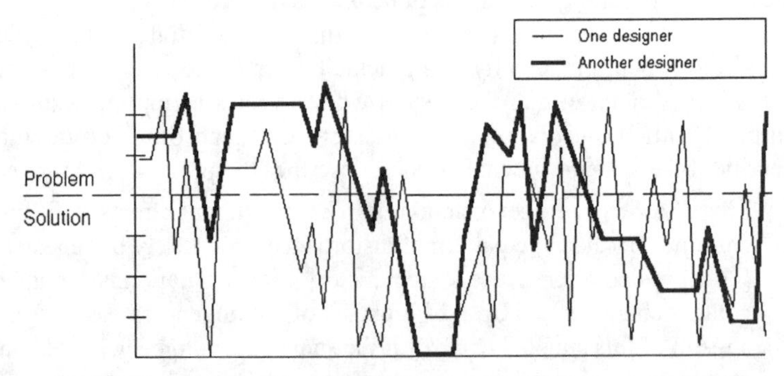


FIG. 2.2. Lack of predictability in how a person works at a problem. One designer may be working at the overall problem level while another works at developing a detailed solution (adapted from Conklin, 2003).

### Nonlinear breakdowns

Let's look at a few examples of nonlinear response within real-world systems.

**Employee communication**
Consider the development of a company that moves from a few people to one that has many. Why does the communication break down rather abruptly? Why does the administrative percentage keep increasing as the company size grow? In the same way, many processes that work well for small examples or small groups fail miserable when scaled up to large groups.

**Server performance**
On the hardware side, a server can be an example. Once a certain load hits, the performance starts to drop dramatically. Achieving acceptable performance requires a bigger server or a more efficient operating system.

However, there is a major difference between servers and people. For the same load, the server always performs the same and the performance breakpoint can be predicted. A person interacting with information does display similar performance curve with a breakpoint, but not a predictable breakpoint. The curves for two different people or the same person on different occasions will start to break at different levels.

**Traffic flow**
Consider the traffic flow for a morning commute. Only a few percent change in traffic can result in the speed going from bumper to bumper at 65 miles per hour to stop and roll driving. Yet, if the road is not heavily traveled, a few percent change (even doubling the traffic) does not slow the traffic down. I once had to commute on one stretch of road that was terrible to drive during the winter but in summer, when the traffic dropped since a small percentage of people went of vacation, the traffic as much faster. Yet, visually the traffic volume seemed the same.

## Feedback loops and situation response

The various characteristics of a situation all interact and cause feedback within the situation that causes a response and resulting change of the situation. Understanding and monitoring a situation requires that a person knows the results of the feedback and should be able to anticipate it.

Feedback controls the performance of any system, at least any system that has regulatory properties. The float valve shuts off a toilet because it provides feedback that the tank is full and a checking account balance works as a spending threshold since it gives feedback that the account balance is too low to make a purchase. There are two types of feedback that can occur: positive and negative. Positive feedback reinforces the current action. In many cases, it can be harmful, such as the squealing of a speaker when a microphone gets too close. The sound from the speaker is picked up the microphone, which goes through the speaker, and gets louder and louder. Negative feedback acts to resist or stop the action. The toilet float valve gives negative feedback that shuts off the water at the proper time. Lack of feedback, such as when the float valve breaks, results in the water never shutting off. Positive feedback would increase the flow rate of water into the tank as the tank level increases.

Feedback monitoring must be automatic and rapid. The feedback must be designed to short-circuit cognitive inertia and keep people following or examining the relevant data, rather than rationalizing all results as fitting with the desired outcomes (Marchionini, 1995). Without the feedback, people don't know if the situation is changing as desired. Of course, defining time scales for automatic and rapid feedback depends on the situation. For long term projects, providing the actual versus projected status on a monthly basis might be adequate. In other situations, feedback may require updated information on time scales ranging the gamut of seconds to hours.

Tied with the nonlinear response and dynamic information of complex situations, they often contain unfamiliar or unintended feedback loops that the user does not expect or does not know how to handle. Factors that were not considered important start to exert a major influence on the situation. A major problem with unfamiliar or unintended feedback loops is that it can be difficult for a person to understand what is happening in time to keep the situation under control. Especially when a positive feedback loop occurs, actions that would have worked without that loop might, in fact, aggravate it.

Person need to monitor a situation to ensure they are achieving their goals. This requires receiving information about the current state of the situation and should include the potential affects of know feedback loops. Because feedback provides the control of how the situation develops, the user must understand its affects to know if the entire situation is currently moving toward the expected state. With slow feedback, poor decisions can have major effects before the person even realizes a modification is necessary. How would you like to drive a car if after pressing the accelerator, you had to wait for 15 seconds before anything happened? Or worse, it only responded to the current accelerator position every 15 seconds and ignored everything in between? Feedback shows up on business reports as the percentage difference between expected and actual values. Poor report feedback happens when making the percentage difference comparison means the person must look at a different report (perhaps last months) and do the calculation manually. Looking at two reports imposes cognitive overhead and complexity which could be easily avoided with better design.

### Feedback for tracking situation development

As part of understanding a situation, the user may make decision that result in adjusting or changing the situation. How those decisions are effecting the system must be tracked. Within the system supporting the complex situation, feedback between the reader, the information creator, and the designer must occur for the system to enhance its support of understanding the complex situation. Feedback provides the basic information that lets the user know if the entire system is currently following the expected results.

Achieving a goal requires that the user track the evolving state of the situation so they can see how the situation is responding. In other words, the information system must provide data so that the person becomes part of the feedback control. Some of the major reasons high quality and timely feedback are required are:

- People often get distracted by the lack of predictability in the complex situation, they need feedback to stay focused on important criteria (Andriole & Adelman, 1995, p. 26). Improper or confusing feedback can lead to people focusing on the wrong data.
- Complex situations often contain unfamiliar or unintended feedback loops (Reason 1990, p. 177). When an unexpected feedback loop starts to exert a major effect on a situation, the user needs to know because it results in unexpected responses from the overall situation.
- In systems with no or slow feedback, people are unable to make mid-course corrections because they don't know the current status.

## Multiple dimensions of a user's understanding

Most single sourcing projects currently develop multiple static documents from a common set of documents or from a text database based on templates. But supporting complex situations requires dynamic documents that get created on the fly with custom information based on a user profile or user selected options. The creation of dynamic documents which support all the dimensions of a multidimensional audience analysis requires a textual database, but goes beyond populating static templates to allowing dynamic expansion and contraction of the information to fit the reader needs. Quite simply, dynamic documents that support complex situations are not just static template-based documents, but ones that are dynamically built to conform to each reader's knowledge level, desired detail level, cognitive ability, and any other situation-specific aspect the audience analysis has deemed important for that particular user group (Albers, 2003b).

Most explanations of how to perform an analysis operate in a single dimension which results in an unordered linear flow of data about the reader or task. All points that need to be considered are listed without considering of how they are interrelated or how they affect each other. The difficulty of both collecting the data and performing an analysis that reveals information relationships prevents many writers from using any analysis beyond a superficial level. With the move to dynamic information generation, the designer faces more than just handling large amounts of information, but must consider what factors enter into the reader's mental analysis. The design must dynamically reflect the multiple dimensions needed by people as they integrate and comprehend information.

There are at least three distinct dimensions which must be explored: knowledge level, detail level, and cognitive abilities (Figure 2.3). Depending on the situation, other dimensions may also come into play, with social or cultural factors being common ones. One distinguishing element of these dimensions is that they are orthogonal, which means they are independent of each other. The three dimensions are as follows:

Knowledge — The subject knowledge the user possesses about the topic. This influences word choice and determines how much supporting information must be provided. Much of the current work in audience analysis tries to pinpoint user knowledge level.

Detail — The amount of detail the user wants about the specific situation. This can range from basic explanations to highly detailed explanations about the underlying physical process.

Cognitive — The ability of the reader to comprehend and understand the material. Cognitive ability includes factors such as the person's reading ability, education level, and physical/mental limitations. The text needs to written at the proper level to match cognitive ability.

A major gain for the writer using a multidimensional model comes because the areas of interest for specific groups of questions can be defined. Reader's abilities vary along each dimension, a group of readers does not map to a single point, but rather to an area (technically a volume). The analysis can define the expected high and low values for knowledge level, details, and cognitive ability. Then for specific reader goals or information needs, the area of interest can be mapped out. The writer's task becomes one of developing content which provides clear communication throughout the shaded area. As an added gain, rather than having to work in terms of expert/novice or other classification, the different user groups and their needs will emerge as part of the analysis.

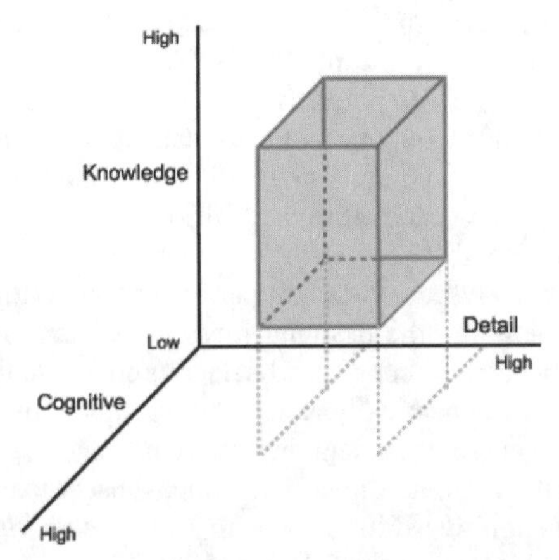


FIG. 2.3. Information areas mapped along multiple dimensions. A specific information need maps out a shaded area; the dynamic content generation must match the text to a particular reader's location within that shaded area. In most realistic situations, more than one shaded area would arise from the analysis.

Figure 2.3 contains a shaded areas which provide an example for specific reader needs. Although Figure 2.3 shows the shaded areas as cubes, in reality the actual shape can vary to fit the information requirements. I can easily envision times when it would be more of a squished football ball shape or a pyramid with the top point being the high knowledge and consistent detail level and expanding as the audience has less knowledge and wants varying detail levels. Figure 2.3 provides a good visual of why dynamic content generation will be required to effectively address this issue. The actual text appropriate for the top and bottom limits of a shaded area may be very different. On the other hand, this mapping is part of a qualitative analysis, not a quantitative one. The collected data does not allow for a concise plot of points that perfectly define a shape, nor do they allow for a precise and meaningful value to be assigned to each axis. But the overall plot does provide a clear visual of what information must be provided and what the user expects to gain from it.

## What it means to understand a complex situation

Having discussed the six characteristics of a complex situation, we will now look at what it means to understand a complex situation. For simple situations, understanding often means no more than knowing a task needs to be done and being able to do it, whether a procedure or look up a value. However, for complex situations, the open-ended nature of goals and information needs prevents complete answers and, consequently, a clear way of predefining when a person understands a situation. Instead, understanding comes when a person has sufficient information to act on it in they way they see fit. In other words, for some people a very high level understanding is sufficient. As part of gaining that understanding, the person iteratively interacts with both the situation and the information system. We see that understanding comes from adequately addressing the user goals and gaining an awareness of the main factors influencing the situation and how it might change in the future.

Understanding comes when the user interprets the information with the proper mental model, which means the system must present integrated information that matches the goals and information needs of the user mental model as closely as possible. In a properly designed information system, users can rapidly access highly integrated information, work in an environment that allows for dynamic modification of data format, and receive support for examining related information, or cause and effect relationships. Understanding a situation via interacting with an information system requires a design focused on achieving information coherence by revealing the relationships across all the major pieces and lets the users synthesize the work as a whole (Thuring et al., 1995).

In the "old days" (modern versions store the same report in an online file), each month the corporate computer department printed out thick stacks of reports on green bar paper. The recipients of the report was then expected to sort out the data by picking and choosing what was relevant, combine it with data on other reports, and make sense of it themselves. This method of dumping data on people and hoping they can sort it out (the basic idea behind current Internet search engines) fails because it does not contribute to letting the person get a grip on the information so they can turn it into knowledge. The weakness of this style of delivery is that rather than providing information that directly addresses a person's goals, the person is provided nonintegrated data which fails to account for the dynamic, interactive potential of the online environment (Heba, 1997).

## *Understanding a situation and situation awareness*

Understanding a situation requires mentally integrating many pieces of information with respect to the user goals and the current context. It requires understanding both that the information exists and how it is interrelated to the situational context and other pieces of information. "Klein stated that a desired theory of situation awareness should explain dynamic goal selection, attention to appropriate critical cues, expectancies regarding future states of the situation, and the tie between situation awareness and typical actions" (Endsley, 1995, p. 34).

The first step in the process of design that supports complex situations is to help the users identify the important elements of the situation and the relationship between the elements. Situation awareness research has focused on understanding these elements and relationships. This section briefly discusses situation awareness research, which states that understanding a situation has three levels which progress from simply perceiving the data to using it to make future predictions, and then considers how it can be applied to the situations described in this book.

Situation awareness was developed primarily within the aviation industry and has also been applied to areas such as plant process control, thus its research deals with short term, tactical situations and decisions (Endsley, 1995; Takahaski et al., 1997). For example, typical research studies include how military fighter pilots can more effectively remain aware of their aircraft and enemy aircrafts while engaged in rapid maneuvers, how air traffic controllers can maintain a clear mental picture of the airspace around a major airport, and how nuclear plant operators can keep a clear picture of the plant conditions during a casualty. In general, situations of interest to situation awareness research require a fast, correct decision under short time constraints and potentially catastrophic or very expensive results from an incorrect decision. On the other hand, the complex situations addressed in this book focuses on longer-term information collecting and analyzing and normally lacks a risk of catastrophe. However, the basic ideas of situation awareness apply to any problem-solving situation (Randel & Pugh, 1996), which includes the complex situations described in this book. Because these differences, I choose not to use the term situation awareness and instead use situational context; however, the three phases of situation awareness described next are relevant. The three phases provide a foundation for understanding the user's goals and information needs and structuring the analysis results for presentation from the information system. How

## Different views of understanding a situation

Consider the different viewpoints of a person coming into a library and the reference librarian with respect to the same situation.

A person stops at the reference desk in a library and asks for help finding material for a medical research project. The library works to gain an understanding of the situation to help the person. A skilled reference librarian knows that the first user query tends to be incomplete and unformed, but knows how to go about asking questions to bring out what is really wanted. What specific area within medicine? How much detail is needed? Finally, having gained an adequate understanding of the situation, the librarian helps the user find the information and the user walks off with a big stack of books and a list of journals articles.

Now, from the user's viewpoint, the task of understanding is just starting: The information has to be read, integrated, and understood. A much more daunting task than collecting the information. (Not unlike having completed collecting information for the literature review of a PhD dissertation. The information is in this big stack, all that's left is putting it onto paper.) The person has the information, but now they must read and understand it in context to continue the research project.

to accomplish that analysis and structuring requires much research to determine how to apply the concepts to dynamic, web-based informational systems.

Endsley (1995) defines situation awareness as more than just awareness of numerous pieces of data. It also requires an advanced understanding of the situation and projection into the future, based on the user goals. She described it as having three levels:

- Perception of the elements in the environment
- Comprehension of the current situation
- Projection into the future

These three levels mean the user understands the information exists, how it is interrelated, and how it affects the future development of the current situation. With poor situation awareness, people can know something is occurring or that a particular piece of information exists, but they cannot easily find relevant information or do not understand how it relates to the overall situation. On the other hand, good situation awareness does not mean a person will make the proper decisions; they may understand a situation, but make incorrect choices.

### Level 1. Perception of the elements in the environment

At the first level, users simply perceive the available information. This level deals with perception, not understanding. Thus, users could have good level 1 situation awareness, but still not have any idea what the information really means or how to act on it.

Users interact with a situation because they want something to change. They do not interact with an information system independently, but instead view it as the best method of answering their question. Belkin (1980, p. 136) discusses how "the concept of information need is the individual's recognition of a problematic situation" and how they fit that recognition into their mental model of the world. To provide good level 1 situational context, the analysis must uncover what environmental or contextual cues bring into focus the problem's existence and what is needed to resolve uncertainty between potential competing problems and solutions. The cueing requires tight integration with the presentation, as users often ignore this step (Hallgren, 1997).

## Reading a medical chart

Consider a non-medical person reviewing a patient's chart (perhaps reading your own chart). You can read the lab and radiology test results, but have no knowledge base on which to build an understanding of your current health situation. One value on the blood work is above the normal range. Should you be concerned and what can you do to lower it? Perhaps, considering your medical problem, that number is normally higher and your physician is happy that its only mildly above the normal range. With your level 1 awareness, you know the data, but have no idea how to interpret it or what to do with it. On the other hand, you expect your physician to make that conversion and tell you about any health problems the lab results might reveal. A major part of a physician's clinical training is learning to make those connections and how to properly act on them.

### Level 2. Comprehension of the current situation

Level 2 situation awareness extends the level 1 concepts from simple awareness of the data to interpreting the information in the current context. The person understands the data and knows how to interpret it within the parameters of the current situation.

Effective support to provide of level 2 situation awareness must explicitly support the mental schema on which users' attempt to interpret the situation (Benyon, 1998). Failure to achieving level 2 situation awareness can result either because the users lack sufficient training to understand the information (which is outside the scope of this book) or because the information presentation fails to either effectively present salient information or to present adequate cues about relevant information. Either case leads a person to interpret the information within the wrong context. Thus, an information system's model of operation must map into the users' current mental schema of the situation context (not the software context). This requires conforming to the users' multi-dimensional strategies for processing information" (Mirel, 1988, 1992).

For most practical purposes, the presentation design of effective informational systems to provide level 2 situation awareness is not a software engineering problems, but is, in fact, rhetorical problems that depend on the information designer. For example, most paper reports provide a collection of separate reports. The problem with these reports is not a lack of information (they usually provide excellent level 1 situation awareness), but they fail to allow for easy level 2 situation awareness. The information is hard to integrate and relate to the current situation. From the psychological standpoint, receiving a collection of reports is highly inefficient and results in high cognitive workload. The reports are not designed for integration; they are designed to present their own information. To use multiple reports, requires retrieving and integrating the relevant information. However, people have a hard time integrating information and relating various data points to each other. Also, they have a hard time remembering or considering subtle cause and effect relations that exist between the information being viewed and other information. Cognitive tunnel vision can cause them to ignore and to forget to consider other information. Years of practice with paper reports have allowed people to develop personal methods of compensating for the integration problems. The moving the information to a computer screen seriously impacted the ability of people to use their learned compensation methods. The result was having to relearn analysis methods and once more exposing the complexities of the data which must be integrated.

## Providing level 2 awareness in a sales report

You review a monthly sales report. The reports contains the monthly sales and percent change from last month. However, these numbers alone tell you very little. How much of the percent change is normal variation, seasonal variation, or the result of an advertising campaign?

Assume that one of the regional monthly sales figures show a drop of 10% from the previous month. To a casual reader, a drop of this size might seem to call for immediate action. However, perhaps these are the figures for January retail sales and the expected decrease from December is 14%. Or that is a northern region dominated by two stores that were closed for several days because of a major snowstorm. Or the advertising flyers mailed late so the extra weekend sales didn't happen. None of this information appears on the sales report, yet making a reasonable decision about the numbers requires the decision maker to know about them. In some shape or form, the mitigating information should appear as part of the presentation.

### Level 3. Projection into the future

Level 3 situation awareness considers the dynamic relationship between elements and how they should change in the near future. The difference between level 2 and level 3 is that level 2 focuses on the current situation and level 3 focuses on being able to predict how the situation will change. In a complex situation, people do not go to an information system simply to find the current state; they want information that can help them understand what is happening now and how the situation will change. With the dynamic nature of the information and the nonlinear response of a complex situation, that prediction often can only go a short time into the future. As such, a person needs to be able to monitor the situation to ensure the proper change is occurring.

## Complex situations are context dependent

To meet their goals, people must acquire a contextually-based awareness of the overall situation. The specific context of the situation plays a major part in how people understand the situation. With complex situations having a history, a person needs more than to know the situation now, rather the history factors must be considered. Of course, the specific history differs for each complex situation, which because of nonlinear response makes each situation unique.

A fundamental design problem that complicates the users' task of understanding the situation arises because many system design and requirement documents lack a clear view of how a user understands the situation. Consequently, the designer lacks a clear sense of what to provide (Gutwin & Greenberg, 1997).

Building the awareness requires communication based on the entire situation and not just the actual text of the message. People come to the information and are influenced in two ways. They carry a large load of social baggage and contextual concerns that cannot be captured by any technological system, but which greatly influence how the information gets viewed and acted on. Also, the rhetorical nature of the content and presentation provides strong influences on the interpretation and human reaction.

The basic problem behind designs which lack the "conceptions of awareness" is because they tends to be ahistorical and generic. By ignoring the importance of situation history and focusing on requirements applicable to all related situations, the design fails to provide the context specific information a person needs to clarify their goals and information needs. The generic design allows the people factors to be ignored, but o–nly at the expense of the overall usability of the design. The interrelations of information within a situation's context must be considered because people are better viewed, not as someone with clear goals (an unstated assumption of a task analysis, but not valid for complex situations), but as someone with fluid, ill-formed goals constantly dealing with ambiguous information (Mumby, 1988) through multidimensional strategies. Bannon (1986) points out that the social structures and organizational contexts are biased toward certain procedures; these biases must be discovered during the analysis phases and allowed for with the design. Schriver (1996) discusses how not understanding the social situation leads to information that fails to address the audience. Failure to allow for the social context consistently appears as part of the reasons that explain systems which function technically but fail because of lack of user acceptance.

A point often forgotten in the drive to make information consistent is that tomorrow the user comes with different goals and so needs a different consistent system (Murray, 1995). A users' context has changed, so they bring a new set of user goals and information needs into what could be viewed as the same generic situation. Part of this contextual change occurs because answering the ill-structured open-ended questions inherent in a complex situation results in what is primarily a cognitive activity situated within a cultural context.

> Cultural context is the mindset that people operate within and that plays a part in everything they do. Issues of cultural context are hard to see because they are not concrete and they are not technical. They are generally not represented in an artifact, written on a wall, or observable in a single action. Instead they are reveled in the language people use to talk about their own job or their relationships with other groups. They are implied by recurring patterns of behavior, nonverbal communications, and attitudes. they are suggested by how people decorate and the posters they put on their walls. (Beyer & Holtzblatt, 1998, p. 108)

Although mention of the word "culture" conjures up images of groups of people closely interacting, the cultural aspects of information are much subtler. As we look at solving users' complex tasks, we discover that the cultural aspects are vital to context (Duin & Hansen, 1996) and that context affects the task. Designing for complex situations requires understanding the cognitive and cultural factors that influence the person attacking the problem. For example, Davis and Flannery (2001) studied how Puerto Rican women perceived sources of health care information and found they disliked receiving it from health care providers and preferred a more culturally sensitive setting, such as a community center. The cultural setting provides expectations and conventions that must be supported for the information to fit users' needs and purposes (Flower, 1994).

As Dourish (1999, p.19) states, it is "essentially an intrinsic element of individual interaction, one which respects that individual action is carried out within a complex of social relationships and provides a means to exploit this connectivity to other individuals to help people organize information." The cultural element pervades all of the evaluation process as information on an individual's desktop is interwoven with the rest of that person's working environment (Chalmers, 1999). Think of the compromises that are made that have nothing to do with the data per se but rather are made because someone will be gone next week, or a senior person always rejects anything with involves a certain factor. For example, last year a new customer relation management (CRM) system was partially installed and then cancelled. As a result, anything that mentions CRM or even might be considered related receives a hostile reception from the management.

Human nature being what it is, people behave differently singularly and in groups and their behavior within a group varies depending on their status within the group. These differences can exert a strong influence on how people understand a situation. As Allen (1996) points out:

> As individuals, people are engaging in individual perception, alternative identification, and alternative selection on an ongoing basis; at the same time, they are being influenced in all of these activities by the social situations in which they are embedded. (p. 88)

On the other hand, there are cultural elements can be considered, but are still extremely difficult to adequately evaluate in advance. The most basic of these is how the people will actually interact with the design and how it will change the current methods. The person (executive level) ordering the work has a specific idea which the system should accomplish. The designers have their own interpretation of the original system goal. But neither know for sure how the real user group will actually use or interact with the system. Unfortunately, the extraneous cultural elements are almost impossible to capture. Both because they are too context specific which cause people to be unconscious of them until invoked, and because, for political reasons, many cannot be clearly put forth in requirement documents. Also many situation contextual factors are one time events, such as rushing a decision because of the need to have something finished before a vacation or before a big corporate event. Such elements are simply impossible to plan for in advance but are a major factor in understanding the situational context in one specific instance.

Although handling of any significant quantity of information is complex in its own right, the organizational constraints and requirements impose another order of complexity which must be considered (Rasmussen et al., 1994). Missing or misunderstanding these organizational constraints can cause technically superior systems to fail. Part of the underlying reasons for user rejection of systems often results when people are forced to work within a technology view that is at odds within their own view of the problem and information space. A problem Dourish (1999) describes as:

> The issue is how the virtual space that the technology creates comes to be peopled and inhabited, how people come to have an understanding of what it does for them and how they should act, and the extent to which the technology lends itself to be taken over by the participants and turned to their own uses. The user must be able to structure the information around their own needs and activities. (p. 27)

Successful design allows users to smoothly work within their view of the situation to meet their information needs and achieve their goals.

## Complacency in thinking we understand the situation

As designers and writers, maybe we're all acquired a false sense of complacency about how to present information within a situation. For the longest time, we could simply focus on our one little piece and ignore the larger context. Think about how the design of a reference manual is to allow a person to look up a piece of minucea. We know that's what the person wants, so we provide it to them. However, put that assumption into context: The person doing the look-up was an expert, so we ignored what may have caused them to want to look up the information. Why did they need to go to a book? What was the reason they needed the information in the first place? Why wasn't it someplace else in the system? If it was, why didn't they look there instead of this place? What is the situation?

Now we're getting someplace. What is the situation? The situation isn't just the system as some designers define it...it's everything that seems to be a part of the real-world position users find themselves in. It has a context and multiple systems often appear as one. Think of your car; how many systems? One or many? Some of you claim it has many, such as: steering, fuel injection, exhaust, and others tremble at the thought of even opening the hood and consider the car, for all practical purposes, one system.

Back to the web for another example. Why haven't we seen the wonderful vision of hypertext realized. Remember the old wonderful claims about how the entire container of human knowledge would be interlinked. click here, then click here, then click here and you could find information on anything. You're reading a history of the theater, click on Shakespeare, click on Julius Caesar, click to read Roman history. In reality, you jump to a page with two items: a link to Amazon to buy the *Decline and Fall of the Roman Empire* and few pictures of ancient ruins in modern Rome. It doesn't work as initially envisioned; never came close. So what happened?

One thing is that the WWW is still basically words with most graphics being eye candy. All words laid out in different formats and writing styles and levels of details. Also, social and political issues arise. Some corporate web sites mandate no links to off-site web pages; for marketing reasons, they try to capture users and not let them leave. Of course, corporate web sites are not interesting in providing world knowledge, they are concerned with selling products.

And that brings us back to the us as the author. What does the reader want, when do they want it, and why do the want it. High quality information design must answer all those questions. A beautiful design of a bus schedule or list of museums in a city isn't enough any more. The two should be smoothly integrated. As I leave my hotel, which bus at what times will get me to and from a Roman artifacts exhibit on display this month. And are there any good places to eat after I look at the exhibit but before I go back to the hotel. The user goals and information needs can now be addressed and can revolve around the online bus schedule. But those requirements are much different from the design of a gray sheet of paper hanging at a bus stop. It's a whole lot harder to effectively design.

I'm willing to claim that we are not even close to being able to properly design it right now or to support creating and dynamically generating the content. Web development tools are only beginning to provide support for dynamic information, very few designers look at design concept from that big of a picture (or have the time and budget to. But that's another issue.), the delivery methods barely support it (Morville, 2001), and the maintenance superstructure doesn't exist. Will it exist someday? Morville is convinced it will and this book is looking at one way to achieve it. But I do know that if we continue to follow the technology-driven road we're currently on, that we're not going to get there. The presentation and technology will never become transparent leaving only information just as the reader wants it.

We've got a long path ahead of us to before users take dynamic integrated information on the Web for granted. And if we're not careful, we'll never get there; the technology will crunch us all.

# Long example—Electronic medical patient record system

Electronic medical patient record systems are a current growth area in health care as hospitals strive to decrease the amount of paper they must maintain and also to better monitor patient care. In general, the system design and research seems to be focused on how to convert the current paper record directly over to an electronic record. But the loss over access control (and having to learn a new way) by the physician seems to an overarching problem inhibiting implementations. As part of the design, a focus is on capturing and providing access to all the information, rather than a focus on seeing the physicians or nurses goals and information needs as a complex situation.

A few years ago, I worked for a company that sold an electronic medical patient record system. It was designed for hospitals to put on each floor (maybe in each room) and to store the entire patient medical record. This system replaces the existing paper charts that are maintained on each patient. The doctors and nurses could log on and view the information they needed to track the patient's progress. Each hospital department (e.g., radiology, physical therapy, laboratory work) entered their work into different systems which then transmitted it to the patient record system (surprisingly, for me at least, the data transmission between these system from different vendors worked quite well).

## *Patient medical records as a complex situation*

If a patient medical record is a complex situations, it must fit the characteristics of a complex situation.

No single answer

Most medical situations can be handled in multiple ways, all which can be considered correct. This aspect becomes even more pronounced as the physician considers the interactions of related diseases or conditions of the patient.

The lack of a single answer especially applies to medical areas that are a hot research area or if the physician needs and wants to keep up on current research. Then, as new research is published, the baseline information, and consequently the acceptable answer, can change. Of course, physicians rarely change their medical decision based on one research study, but they need to be given information to build their own answer.

Open-ended questions

Besides lacking a single answer, the amount of information a physician or nurse wants varies. Likewise, how detailed of information they want varies depending on their position and background.

| | |
|---|---|
| Multidimensional approach | The actually approach taken to understanding the situation varies based on the person. Consulting physicians will review the information differently than the primary physician as they have a different level of interest and focus. |
| Dynamic information | Adjusting for the various approaches and detail levels means the information must be dynamic. Likewise for meeting the different knowledge levels of the audience. |
| History | The information a healthcare provider needs depends on the past history of the patient; patients with chronic problems can greatly affect the available treatment choices and rate of recovery. |
| Non-linear | The amount of information a person wants can shift abruptly as they learn more, as the disease progresses, or their responsibilities change. For example, a person might be content with a general level of knowledge a parent's disease until they are suddenly tasked with being the primary caregiver. They may also want a different view of information is it inspires issues such as "this part sounds like what my neighbor did before he went to a home. Is Mom going to have those same problems?" The two diseases might be different, but the person needs information to distinguish them. |

### *Human-computer interaction*

The program's interface for information retrieval, which a physician would use when checking on patient status, was the exact opposite of the integrated information presentation I've discussed in this book (Figure 2.4). The top part of the screen was patient name and such. In the middle was a multiple rows of tabs with names for area that might enter information into the record (physical therapy, radiology, urinalysis, etc.); each area of the hospital had its own tab. Clicking on the tab brought up a chronological list of the information. So, the most recent blood test result was first and the doctor could scroll down to see yesterdays and the day before.

The fundamental issue here is that a set of tabs is a common interface design method when many different items needs to be easily accessible. However, tabs assume that the information behind each tab are independent. When that assumption is true, tabs are very useful. The tabs at the top of Amazon's homepage work because the user goals fit that assumption. But in a medical implementation, the information is all interconnected; a basic usability assumption of tab design is violated.

In another common usability design problem, notice how the system treats each piece of information separately and of equal value. If you ever have a chance to watch a doctor on rounds looking at a paper chart, you'll see three or four fingers marking pages, and constant page flipping back and forth to compare values. With the tabbed interface, there was no easy way to compare a value on the urinalysis with a value on a blood test. Even worse was trying to figure out how those two values had changed over the last couple of days. The values could be graphed, but graphing was considered a separate feature that was not connected to the currently viewed tab. After opening the graph window, the information to be graphed had to be pick from a long pull-down list.

As a secondary usability issue, but one critical to mentally integrating information, was the response rate. When a tab was clicked, the system required 3 to 7 seconds to retrieve the information and present it. Yes, to look at test result A, compare it to B, and look back at A could take 5-15 seconds. With a paper system, a physician could make this check in only a few seconds. Besides being generally infuriating for the user, this slow of a response effects how they use the system. Wait times which are that long can result in people forgetting the value of result A (short term memory decay). They also will not look at result B if they feel pressed for time or feel reasonably confident in result A, a check which would have been made if the response was faster. In other words, they don't perform as deep as analysis. A computerized system must provide time responses at least equal to a paper system or must provide a strong benefit in other categories. Although hospital administration might see strong benefits, to the medical staff actually using he system, response time is a major factor to acceptance.

These issues of use of tabs, treating each information element as of equal importance, and slow system response result from considering the overall situation as a simple rather than a complex one. The users don't need the value of one piece of data, rather they need an integrated collection that must be combined with human problem-solving to arrive at the proper decisions, either the initial diagnosis or decisions about continuing care.

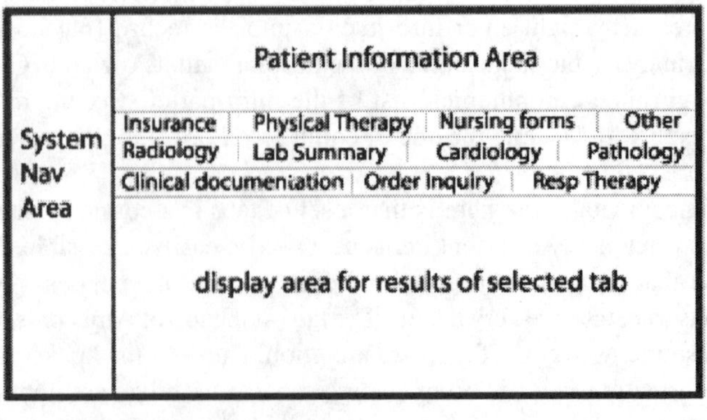

Patient Information Area

System Nav Area

Insurance | Physical Therapy | Nursing forms | Other
Radiology | Lab Summary | Cardiology | Pathology
Clinical documeniation | Order Inquiry | Resp Therapy

display area for results of selected tab

FIG. 2.4. Mockup of electronic patient record screen.

The company had already shipped the initial version product, so was not interested in discussing a redesign that considered the complex information needs (which, admittedly, would have been too expensive a retrofit onto the existing interface design). Plus, I faced the old argument of how the customer's was used to this interface so we couldn't change. But consider how the interface could be improved to provide a cleaner presentation of pertinent information. The goal is a system that works within the patient's medical situation to provide for healthcare providers' different information needs.

## *Designer and writer issues*

Chapter 7 lays out a set of steps to consider when designing for a complex situation. This section looks at those steps as they apply to a patient record system.

### Scope

The scope of a patient medical record system can seem rather simple: contain everything about the patient, so it mirrors the existing paper record system. However, as a scope statement that is too general. The individual scope statements must take in to account the limitations of both the paper system, so the design can extend or expand them, and it must take into account the limitations of an electronic system to ensure the design don't inhibit the user.

Some sample scope statements are as follows:

- Track recovery progress against evidence-based expected response.
- Focus on test results pertinent to the diagnosis.
- Ensure care for multiple health problems matches evidence-based medicine recommendations.

### User groups

Nurses and doctors have different information needs. Is the doctor on rounds, concerned about a complication, checking if the patient can have a test or receive a particular medication? Is the doctor a specialist called in to consult on this case? A nurse's information needs vary by the current patient task. Likewise, the other medical personnel, such as respiratory or physical therapists, each have different information needs. Each of these requires different information since they have different goals. Each of these have to be addressed during the design and analysis phases.

### User goals

A medical patient record system must contain all the information a physician needs about the patient, but even assuming the diagnoses exists, the exact information and sequence which a doctor desires the information remains unknown. The analysis needs to start with the different times that the system gets accesses and by who.

### Information needs

The information the healthcare provider really cares about varies depending on the diagnosis. For example, both the specific items from the lab tests and how they are compared with other lab tests can change. Will this greatly complicate the design analysis? Yes. The designers will need to work through lists of diagnosis one at a time and capture what the medical practitioner considers more salient, how it should be best viewed, and what should trigger possible warnings of complications.

### Information relationships

None of the individual information elements are difficult to find and there are not even that many of them. Yet, making the connections and understanding the relationships have stymied expert system design. Almost the entire complexity, and the reason physicians spend so many years in training, stem from comprehending the relationships. Thus, the design issue is not to try make the diagnosis, but rather to provide the information arranged in a manner to allow a physician or nurse to make the proper decision quickly.

An interesting thought here: we expect and legally demand that a physician spend years internalizing this information, yet it's considered too complex and time consuming to capture during a design analysis. Not to mention that medicine has developed a culture that looks down on having to look up information versus having it all stored in the physician's mind.

### Source

Depending on the overall system design, the information sources can be the easiest part of the system design. There are many vendors that specialize in data entry for various departments and that support data exchange with other systems. The data source becomes a technology problem rather than a content management problem.

### Presentation

The presentation is critical since different presentations can strongly influence a person's decision making process. Once a diagnoses is made, the system should present the information in a form that fits the diagnoses. What test values should be compared and how should they be compared? What tests are recommended? Access to medical textbook information about the diagnoses can also be available. Rather than display all lab results, only initially show the ones pertinent to the diagnosis. Of course, the full results of the blood work must be easily available, but if only certain values are normally looked at for this diagnosis, then the initial presentation can restrict it to those values and display them in an appropriate way. Perhaps shown as a graph of values over the last 24 hours, with a second line of a different test result that should track the first.

# User Goals 3

*Our wants are infinitely beyond our needs, and our expectations about the world almost always err on the side of optimism.*
*—Gary Marchionini*

*Making the complex clear always helps people work smarter. Because it is a lot easier to figure out what's important and ignore what isn't.*
*—Bill Jensen,* Simplicity

Chapter 2 examined complex situations. This chapter looks at goals and how they form the most important aspect of a person working in a complex situation. In looking at goals, this chapter:

- Defines what makes up a goal.
- Considers goals in complex situations and how they relate to information needs.
- Looks at the various constraints that a situation can impose on a user defining and achieving a goal.
- Discusses how in complex situations goals emerge from the situation itself and cannot be predefined in detail.
- It looks at ways of ensuring a design helps a user set and achieve the goals.

## Introduction

In chapter 1, goals were defined as "The real-world change of the current situation or understanding of the current situation which the user is trying to achieve." This section expands on that definition to clarify what a goal and sub-goal means, and considers what changing or understanding a situation means in the context of a user and a situation.

The definition of a goal talks about changing or understanding a situation. Some changes to a situation are things that can be measured: increased employee retention, airline tickets are purchased, or the form has been emailed. Other changes are cognitive. Understanding is a cognitive change in the situation; before the user didn't know which dive boats were available and now she does. Now she can begin to find information about how the dive boats differ and evaluate how those changes might effect her trip. With cognitive understanding, although from an external view the situation may be unchanged, the user's view of the situation has changed, thus the way information gets interpreted has changed.

## *Goals and user groups*

A situation always exists with the user embedded within it. In the model for complex situations used in this book, achieving a goal means the situation has successfully transitioned from one state to another. Making that transition means the user needs to either understand the current situation or modify it in some way. To accomplish either the understanding or modification, the user works within the situation to set up goals and searches for the information required to achieve the goals. For understanding, the information may directly map to the goal, whereas for modification, the information provides cues about how to modify the situation to achieve the goal.

People set goals to guide them through a situation, but all people are not the same. Different people shape their goals differently and may set completely different goals. Rarely will a complex situation deal with a single homogeneous group of people sharing a common pool of goals. Instead, multiple user groups exists, with each group having a different pool of goals which need to be addressed.

Spool (2003) looked at how people use banking interfaces and found some drove people away by not answering their questions in the user's context. He described the information problem as follows:

> The problem is that team didn't know anything about Leslie [the research participant]. They needed to know why Leslie was coming to the site (her 'intentions'). They also needed to know that she'd just come upon a huge sum of money (her 'context'). The team also needed to know what Leslie knew about investing (her 'knowledge'), what she was capable of doing herself (her 'skills'), and the nature of the financial management she'd done in the past (her 'experience').

Spool then continues to explain what happens much too often:

> The user's intentions, context, knowledge, skills, and experience are the essential things that every designer needs to know. Without this, the team is going to design something that seems useful, but they'll never know if it actually helps the user. The result is exactly what we see with Citibank's design—a lot of content, but not the right content.

A vital part of a design analysis for complex situations is clearly defining who the various user groups are. Then the goals for each user group can be defined. Depending on the user groups, some goals will be identical, some will be very similar with important variations, and some will be unique to the group. For example, when describing product information as part of a potential purchase, both a manager and the senior technical person (the real user) have a goal of understanding the features. But the technical person will want more detail about product operation, whereas the manager will want more detail how it might effect workflow or other big picture aspects.

While acknowledging the importance of understanding who the user groups are, the remainder of this chapter generally ignores them. The users who are discussed here should be considered a member of a specific user group. The nature of user goals is consistent across all user groups; each user group may have different goals, but the reasons to identify them are the same.

## Goals and sub-goals

Any person working in a situation has overarching goals which normally define the desired end point. In a simple situation, the goal can be clearly defined, such as "enter expense report." In more complex situations, they can be "do my taxes," or "find information on Caribbean dive trips." As the situation complexity increases these overarching goals become too general and too vague to clearly define how to accomplish the goal. Planning a SCUBA diving trip has many different parts that each need to be considered: where to go, which airline to fly, which dive operator to use. Achieving the overarching goal requires achieving a host of sub-goals.

Achieving a goal usually requires it to be broken into smaller pieces, which we will term *sub-goals*. (Note that this book normally uses the term *goal* to collectively refer to goals and sub-goals, unless the context requires them to be distinguished.) Sub-goals are normally solved in a recursive manner. Each goal gets broken into a group of sub-goals (which may be further broken down), and each sub-goal must be handled before the goal can be considered achieved. Even with simple situations, such as the expense report, sub-goals exist. One sub-goal might be breaking the expenses into categories such as travel, meals, and car rental. A second sub-goal is entering them into the system. A third sub-goals is getting a manager's signature and sending it to the accounting department. In turn, the third sub-goal contains its own sub-goals: finding the manager, finding and addressing an envelope, and walking to the mail drop.

## User goals and complex situation characteristics

Defining user goals is not a matter of asking users to list all possible goals and then arranging them in order. Rather, it is a matter of understanding what goals the user want to achieve and how the relate to the overall situation. Each of the characteristics of complex situation brings out the importance of this idea.

No single answer — Because of the many different variations in both user goals and potential solutions, a single best solution path simply does not exist (Rasmussen, Pejtersen, & Goodstein, 1994). Instead, the analysis must uncover the cognitive and social aspects which influence the goal setting. Also, while a goal may seem similar, the exact formulation between user groups can differ.

| | |
|---|---|
| Open-ended questions | With no single path, there is no single question a user can ask, nor is there a clear answer to when the question is answered. It can be difficult to define when the goal has been met. For a person seeking information, there is always more information which could be read. |
| Dynamic information | As a person progresses toward achieving a goal, the goal may change or new sub-goals maybe set. Changes in information may force a change in the goal. Goals cannot be preset and the information cannot be prestructured; rather, it must be designed to allow users to continuously adapt the information while searching for a solution (Elam & Mead, 1990). |
| Multi-dimensional | Because users approach achieving the goals differently, their information needs will be different for each approach. Each user group defines one dimension and the different knowledge and detail levels required within each user group shape the multidimensional aspects of the goals. |
| History | The history of the situation and history of external factors affects how people interpret the information and how they will predict the future information values. |
| Nonlinear | Goals can change abruptly as people interact with the situation. They may try to force achieving a goal and then suddenly stop and reform a new goal. |

## Goals in complex situations

Supporting complex situations comes down to meeting the user's goals by supplying users with access to the information they need to assess the current situation and, if required, make reasonable predictions into the future. The interactions between the user and system must provide "sufficient information about current status and activities that, when intervention is necessary, [users] have the knowledge and capacity to perform correctly" (Shneiderman, 1992, p. 210).

The readers bring to the system a set of real-world goals which the system design must consider from the earliest stages (Belkin, 1980). Properly presented information with the proper content effectively addresses the readers' goals (Albers, 1999, 2003). Connecting goals with the requirement of communicating information to the user to provide a clear picture of the situational context means the overall design must support the users' strategies as defined by Marchionini (1995): "Strategies are those sets of ordered tactics that are consciously selected, applied, and monitored to solve an information problem" (p. 73). In the word processing example discussed in the "Goals of a documentation writer" sidebar, we can see that the basic system goals often do not directly map to the user's real-world goals. For a word processing program, this means being able to find information the supports using the program to perform user-defined actions, not simple descriptions of menu options.

Although it can be easy to define a situation, and may even be easy to define the overarching goals, for complex situations, the dynamic, real-world influences come into play and complicate defining the user goals and information needs. When planning the dive trip, a sub-goal "were to stay" might include whether to stay on a live-aboard dive boat or an island hotel. Now, within the "where to stay" sub-goal are more sub-goals that must be addressed: which dive boats are available, what amenities do different boats and hotels offer, how much do they cost. Personal preference factors start to come into play that make prevent straight forward information presentation. What if the person wants to dive lots of different sites or do a specific type of diving? If the person doesn't golf, then a dive resort that give emphasis to its golf course is providing distracting and irrelevant information. These type of factors cause a dynamic influence on the user goals and the information needed to address the goals.

Figure 3.1 shows the results of a system that matches the user goals and one that does not. The matching process is not just one of providing information but one of supporting achieving goals. Many failed system have high quality information, but the lack of connection to user goals rendered this information difficult or impossible to use. In chapter 1, there was an example of a dynamic jigsaw puzzle. The same idea applies to issues in Figure 3.1. As the user interacts with the situation the goals change, which results in the left-hand shape changing. A dynamic system should have the ability to change the shape of the system support to match the changing user needs.

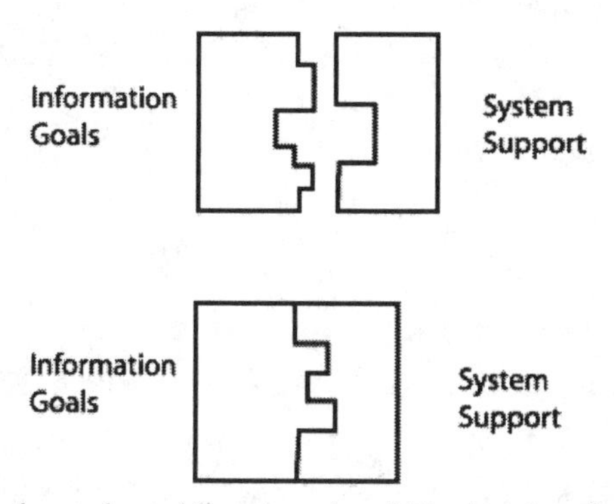


FIG. 3.1. Fitting the real-world user goals within the situation to the goal-information presentation of the system. Only when the two match can the system effectively communicate the information the user needs to confidently achieve the goal. (based on Vicente, 1999a, p. 39)

Of course, depending on the situation, the design can be rigid or highly flexible; in most complex situations, a highly flexible strategy is desired. A strategy based around "whether the information will be logical to the user in the context of the use of the information" (Duffy, 1995, p. 269) and will it help achieve the goal is preferred.

## Goals of a documentation writer versus goals of a word processor user

A writer engaged in documenting a software product normally focuses on explaining each function and menu option. The goal is to have a complete description of the function. However, the goal of the user is not to obtain a complete description but to apply that function to a current situation.

Consider a word processor user who needs to create a page with an image in the running header and left aligned text, like this:

Click

Photography
Methods

The user goal is to create a page that contains that running header. Nothing in the documentation will explain how to do that. The documentation will explain running headers, but may or may not mention that graphics can be included. The section on running headers will also not mention how to get the right aligned text. The section on paragraph formatting will not explain how to use tabs to get the text into the desired format. Nothing will tell the user to draw the line with the drawing tools instead of using borders and shading (in Word).

The documentation will contain all the required information about inserting graphics, setting tabs, drawing lines, and aligning text, but it will not be arranged to fit the user's goals. Granted, in a paper document, it simply can't. But with dynamic content generation, if the user goals are understood, then specific custom text can be created.

Each of the ideas are actions a person might want to do with Word:

- Reversed out text (white text on black background)
- Graphics in the middle of two columns
- Heavy rule above a heading and only the length of the text

Accomplishing each of these requires the use of multiple options. None are explained in most books on Word. In paper, the various options are too extensive and would create a excessively large document, but it could be done with a web-based dynamic system. The analysis would have to define the user actions (such as create reversed out text) and then develop the appropriate content to support it.

### *Relationships between goals*

Connecting different goals and their associated information needs are their relationships which contribute to the multiple dimensions of a complex situation. In effective design for complex situations, everything must focus on providing effective answers to those real-world questions. For a designer to provide support for these goals requires understanding both what the situation is and what it means to the person: How does the person view the situation, what previous knowledge do they bring, and what type of transitions might they want to accomplish? Achieving the design goal of supporting the user in understanding the relationships between goals involves:

- Finding the logical relationships between different user goals and available information.
- Defining the information needs to make those relationships both easily visible and clearly understood.
- Emphasizing why the goals are as connected and how that connection relates to the overall situation.
- Defining how the social and cognitive aspects interactions inherent in the situation affect both goal relationships.
- Showing why or when the connections between goals are relevant or when they are irrelevant (with the latter being just as important as the former).

### *Goals versus wants and needs*

Thus far, the chapter has discussed how achieving a goal requires knowing what information is required, understanding the information, and how to manipulate the information to extract the required knowledge from it. The design analysis must work to understand the information needs, how the person uses and understands the information, and the manipulation requirements for the information. Notice that the previous sentences talked of *needs*, not *wants*. Unfortunately, a person's needs and wants are often quite different, with people rarely knowing what they actually need.

Simply asking what they do or need is likely to reveal either the standard line or a view of what the person honestly thinks they do, without noticing the many actions which are carried out without thinking. People don't realize what they do, the performance process too deeply ingrained. They also don't realize everything they need to do it. They often can't even ask for what they need because they may not realize the information exists or may not realize that alternate paths to achieving goal are possible. Or they may be the type of person that does not want to see anything negative or that contradicts their current beliefs; regardless, these people too need to see the information they need. With huge amounts of information available in modern databases, everything is both wheat and chaff, depending on the situational context. The design analysis must define which is wheat (the real need) and which is chaff (unwanted, or wanted but irrelevant) in each situation, to discover "what users really need from software to support their work as distinct from what they want or what they merely think they need" (Constantine & Lockwood, 1999, p. 98). The collected goals and information needs create a vision of the users focused on what real-world questions they want answered and why (Mirel, 1998, 2003).

Since it is difficult to find out what information a user wants, it becomes even more important to understand what a user needs. The importance of a thorough audience analysis and a task analysis cannot be overemphasized (Lillies, 1991; Redish & Schell, 1986; Guillemette, 1989). Audience and task analysis provides an understanding of the reader's prior knowledge, attitudes, and needs, allowing design of an appropriate collection of information. A major reason the audience analysis is so important is that it helps define all the user groups. The audience and task analysis creates a list of the user's needs, which the designer transforms into a usable presentation of information tailored to each group. Information system users span the spectrum from computer novice to expert and from situation novice to expert. Each group's needs are different and, unless they are understand in detail, providing usable information that address the goals of all levels is difficult (Rosenbaum & Walters, 1986). Designers do a great injustice to the user when they refuse to design for each situation and instead either simply provide everything to cover all situations and make the user sort it out, or provide information focused on one situation while ignoring others.

### User goals, not system goals

Besides meeting the user needs rather than user wants, the system must also meet the user goals. Too often, information system designs meet system goals rather than users goals. These system goals include using certain interface design tools which constrain design flexible to creating a design that looks like one the boss likes. They might also include making design decisions based what is easiest or fastest to implement or code, rather than what is best for the user.

At times the situation and system view can be confused by the way the system gets initially ordered or justified. The executive sponsor says they want a system that provides information X, Y, and Z. Obviously, the final system must provide that information, but how it accomplishes it determines how well it addresses the real user's goals and information needs. The sponsors know what information they want presented, but they also often lack a clear understanding of how the information will be used, which may radically change how it should be presented and what other information must be presented. I've seen poorly designed systems that did little more than take the sponsor's list and create a menu option for each one, with no consideration to how those requirements really fit together.

A design requirement may be to present the information about all widgets and allow it to be manipulated. A system-based result has a table of all widgets and pertinent data. This table lists all names alphabetically or grouped under some higher level classification which reflects the organization's production or sales struc-

## Table saws and user goals

Consider how the design of new table saw can differ whether or not the user needs versus the user wants were considered. If user's are asked what they want, the response is likely to be rather vague such as "I need a table saw to cut wood for my building projects." To the user, this may sound concrete, but does not separate out what the user wants in a table saw from what he needs in a table saw.

**Case 1.** Since the user wants to cut wood, the designer decides to see what type of wood might need to be cut. So, he goes to a lumberyard and records the various types of lumber. The resulting saw can handle everything in a lumberyard. It has a huge blade to cut 8-inch beams. The bed can handle 20-foot sections of lumber or 4 x 8 feet sheets of plywood. It fits the description of what the user wanted, although it fails to meet the user's needs because it's too dangerous, big and bulky for the average home-owner. It fails to meet the user goal of cutting wood because it's both too expensive to buy and too large. (Although I do have visions of a *Blondie* comic strip where Dagwood tries to bring this saw into the house.)

**Case 2.** Since the user wants to cut wood, the designer works to uncover exactly what type of wood the user wants to cut (what are the user goals?). Focusing on the real goals revealed rarely cuts more than an 8-foot 2 x 6 and plywood is not cut on a table saw. The resulting saw makes the needed cuts and is light and easy to store.

The difference in the two cases was that the design started with the user's goals rather than the *possible* goals (all the wood in the lumberyard). However, the final homeowner design will not cut all the possible lumber. It can't cut an 8-inch beam, but if the user never wants to make that cut, it doesn't matter.A different design will be required for the user group that has to cut 8-inch beams and a 20-foot 2 x 12 because the fundamental needs of these groups differ dramatically and can never be addressed adequately with one design. Clearly defining the user goals also helps prevent design-creep during development by shutting down the discussion as irrelevant when someone points out that the saw will not cut an 8 inch beam.

Although some may consider this an absurd example, the absurdity stems from a table saw being a physical object. It seems totally obvious that everyone in the world should not use the same saw. Yet when it comes to information systems, we have a false idea that it can be infinitely malleable and can handle all situations.

Quality design requires working from the user view, not the technology view. Many of the real difficulties in designing an informational system do not fall within the technical realm. Rather, they fall within the user realm. The technical realm addresses solving problems by placing widgets on the interface and developing algorithms for an efficient (from the software engineering perspective) back-end. The user realm addresses situations by developing goals and information needs which conform to user goals and the organizational context into which the situation exists as a single element of a much larger puzzle.

ture. By clicking on the column headers, the table can be sorted according to the column. Ordering by columns is easy for a developer to provide, but in complex situations user goals rarely require a simple ordering. Normally, the user only cares about a few widgets and either needs more or less information about them than is contained in a table. Instead of always having to deal with a table with all widgets, once the user selects some subset, the other widget information should disappear. Also, if specific information, which may be too technical or situation-specific to fit into the all-widget-table, fits the current situation, then that information needs to be presented. Marchionini (1995) describes the fundamental problem with this widget table when he wrote:

> A key problem in information-seeking performance and information system design is clarifying different levels of the task's goals. For example, the goal from the system's point of view is to provide a document of some sort, whereas the goal form the information seeker's point of view is to extract information (make meaning) from some document and stabilize or advance his or her knowledge state (p. 37).

Providing the level of design required for complex situations requires that goals be defined and considered from the situation viewpoint, (what is happening to real users, what do they need and why) rather than the system viewpoint (how can the system display value for x). Creating usable dynamic information systems requires understanding the questions going through the users' minds and what information they need to construct a clear picture of the situation. It then requires a design which contains the contextual knowledge to filter the data and create an integrated presentation (Bowie, 1996). Providing this contextual knowledge means designing from a detailed understanding of the user environment and user goals, not the system goals.

Consider the difference between user and system goals which Marchionini refers to. In the example on widget information in a table, the design failed to address the user goals and focused on the system goals. Answering the user's real-world questions requires information designed to fulfill their information needs (Albers, 1997, 2003). Addressing the user goal means the design must allow for manipulate the document order and content to extracting the user-defined meaning and relationships. Cohen (1993) addresses the design problem by saying that rather than forcing users into solving problems the computer's way, the designer must "start with the user's preferred way of solving the problem and to examine its strengths and weaknesses carefully" (p. 266). Cohen's statement supports Mirel's (1996) claim that "the most important factor in designing effectual data reports—even more important than the format of a table—is for writers to select and present only the information integral to their communication situation" (p. 95). Rather than taking an attitude of "give it to them and let them massage it," the effective design must work from the user goals to define what information is needed and when it is needed, and then provide that information to the reader in an integrated format that fits the specifics of the current complex situation.

## *User goals, not user tasks*

Task analysis attempts to capture the application information at the level of individual user task and attempts to draw out the procedural knowledge inherent in the task. Most task analysis methods are designed to analyze well-defined problems via a rigorous analytical method. Task analysis basically assumes a linear flow of

## Question and answer format

Using a question and answer format has become a common method of providing information.

Internet FAQs are one example of a question and answer format. Most Internet FAQ lists are system focused rather than user focused. It seems as if the questions were constructed not from real user questions, but by someone sitting down and thinking "what might someone want to know?" or worse "what do I want them to know?" Consider that the information gets presented in a question and answer format with the questions implicitly posed as "what does function x do?" or "how can I do x?"

Web sites that have huge and complex information source also tend to use question and answer format (a medical example is MedicineNet at http://www.medicinenet.com/). It also seems users are more prone to using discussion boards and other methods of connecting with people rather than going to online information and figuring it out themselves (Wilson, 2003). The user navigates through a hierarchy to information on a particular disease and eventually gets to a list questions. Each question has a short paragraph answering it. If the reader want more information about the answer, doesn't understand some of the terminology, or generally has trouble understanding the answer, too bad. With a dynamic system, the reader could manipulate the answer to get more detail, have it provided at a higher knowledge level, or change the reading level of the text. But even more, rather than just getting an answer to one question, a dynamic system could fit the reader's needs and connect that answer to the entire body of knowledge on that disease.

The information provided as the answer for each question is normally correct. There is not a problem with the writer intentionally trying to mislead the reader or providing incorrect information, rather it's a question of whether readers can actually achieve their information gathering goals by reading the answers. If mentally forming a clear answer requires reading and integrating the information of two or more different questions and answers, many readers will be lost because the connection is not explicitly made. For the ill-structured tasks inherent in complex situations, rather than just presenting questions and answers, the information designer must anticipate the interrelations of the information and provide the information when users want it. These would include attempt to address goals that involve using functions x, y, and z, something missing from question and answer systems (and the vast majority of software documentation).

actions; a sequences of tasks are performed which take a system from point A to point B. With an emphasis on functional decomposition, task-analysis methodologies tend to be product-oriented and do not pay sufficient attention to the nature of the socio-cognitive environment in which systems are implemented. (Lansdale & Ormerod, 1994).

Part of the reason task descriptions focus on system behavior is they often become an end rather than a means isolated from any higher purpose or social context (Jirotka & Goguen, 1994), suffering a lack of connection to the real-world task users want to accomplish. Although the longer discussions of task analysis (such as Hackos & Redish, 1998) do explain the necessity of capturing human behavior and why the task is being performed, the concise "6 steps to performing a task analysis" article common in material aimed at practicing information designers ignore that aspect. These short articles normally use data entry type systems, a simple situation that eliminates the need for extensive human behavior analysis.

Whether by direct observation, interviews, or surveys, all tasks and their order of performance are assumed to be observable in a tangible way, an assumption Mirel (1993) found overly simplistic and one that misses the underlying reasons that motivated people to perform non-trivial tasks. Across any analysis of all the users, designers need to ensure that the analysis has not revealed just one method of many that are used. Forcing users to perform the task in a different way interferes with their work patterns. Users insist on performing tasks in a manner that fits the context. However, when the task analysis divorces the task from the real-world context, it loses the knowledge of how the task actually fits into the users' world view (Waern, 1989). This often fails to provide an understanding of the content needed to make the system work. The analysis describes the events performed by the system, but do not "explicitly define users' information requirements" nor provide insight into to how to most effectively perform the action (Sutcliffe, 1997, p. 223).

I am not opposed to task analysis; a task analysis works wonderfully and provides the analysis information required when it is used appropriate situations. Traditional methods of task analysis are suited to describing computer interaction as discrete events, but are less useful for the "continuous, holistic process" (Basden, Brown, & Tetlow, 1996, p. 157) of complex situations. For routine, well-structured tasks, the assumption task analysis can uncover explicit procedures holds true. For the well-structured problem, task analysis methods work and should be used. For example, if the task analysis shows that only simply, straightforward tasks are performed, the design considerations which are invaluable for complex situations will not necessarily apply. Conventional task analysis works well for routine tasks, but routine tasks are not within the realm of complex situations. Actually, for a simple situation, a task analysis would provide the desired information and the user goals and information needs approach of this book would probably provide minimal benefit and greatly increase analysis complexity. On the other hand, as Cohen (1993) explains:

> Computer-based systems to advise decision makers have incorporated decision analysis, expert knowledge, and/or mathematical optimization. Success, however, has been limited; the very features of real-world environments..., for example, their ill-structuredness, uncertainty, shifting goals, dynamic evolution, time stress, multiple players, and so on typically defeat the kinds of static, bounded models provided by all three technologies. Each decision involves a unique and complex combination of factors, which seldom fits easily into a standard decision analytic template, a previously collected body of expert knowledge, or a predefined set of linear constraints. (p. 265)

In analysis for complex situations, rather than thinking in terms of tasks, the analysis must proceed in terms of user goals and information needs of the situation as viewed within the user's real-world context (Albers, 1996, 1999, 2003). Rather than trying to have the system handle the situation, we must take the approach of the system helping users identify and clarify their goals and information needs, and having users decide on appropriate paths (Belkin, 1980). For complex situations, analysis must work to "define the problem you intend to solve in terms of the work you plan to support" (Beyer & Holtzblatt, 1998, p. 67). The issues are not data handling or button pushing, but people handling. "Ask: What is the work we expect to support? How does this work fit into the customer's whole work life? What are the key work tasks? Who is involved in making the work happen? Who are the informal helpers? Who provides the information needed to do the job and who uses the results?" (Beyer & Holtzblatt, 1998, p. 68).

These contextual design ideas represent a shift from thinking in terms of tasks to thinking in terms of the user's mental process. In other words, a shift from a strict task analysis to an analysis of the user's goals and needs as they apply to the task and an objective of working at the same level at which people actually work and the level at which the task has contextual meaning. A level that Carroll and Rosson (1996) described by saying "We are concerned with tasks at the level people construe their work to themselves; the level at which tasks become meaningful to the people who engage in them. Many task analysis schemes focus on a much lower level than this" (p. 231). A analysis for complex situations represents a shift away from the low level processes captured by most task analysis and a shift toward placing the captured data within the context of the user's real view of the situation. With a higher level of focus than technology-centered methods, user-centered methods provide a better, although much harder to perform, approach to addressing complex problems (Butler, Esposito, & Klawitter, 1997).

## Constraints on achieving goals

The constraints imposed on accomplishing a goal can be split into three major categories. The first two, cognitive and environment, were discussed by Vicente (1999a) with regards to accomplish a task. With complex situations, especially the cognitive ones based around gaining information, technology also provides a constraint.

| | |
|---|---|
| Cognitive constraints | Constraints which originate within the human cognitive system. This includes limitations on reading speed, memory, and the ability to move between disparate information sources and integrate them. |
| Environment constraints | Constraints which originate within the situational context that make up the physical and society reality in which the situation is located. This includes factors as varied as the business rules and how much time the boss allows to analyze the information, and the broken air conditioning system. |
| Technology constraints | Constraints which originate within the technology used to collect the information from the situation or to present the information to the user. This includes factors such as the size and clarity of the computer monitor, and how the database handles information. |

### *Cognitive constraints*

Cognitive issues focus on the abilities and limitations of human mental processing. This includes limitations on reading speed, memory, and the ability to integrate disparate information sources and them. Cognitive issues and how they effect complex situations make up the bulk of chapter 5. This section only touches on cognitive issues that work as constraints on achieving goals since they will be addressed greater detail in chapter 5.

When working with complex situations, the user has goals which normally require the evaluation of information from multiple sources. Also, as discussed earlier in this chapter, rather than needing information in pre-defined ways, the viewing order and specific information required changes with each view of the situation or goal. Cognitive constraints or problems include:

- People find it difficult to effectively search for and integrate multiple sources of information (Lansdale & Ormerod, 1994).
- How information gets presented can cause people to make opposite choices and believe them to be best (Tversky & Kahneman, 1981; Johnson, Payne, & Bettman, 1988). In other words, using the same data with one presentation people pick option A and with a different presentation pick option B, and feel they have made the correct choice.
- Expectancy bias (seeing what you expect to see rather than what is there) comes into play and can only be seen clearly in hindsight. Expectancy bias helps explain why in many disaster or major accident situations, people made seeming bad choices and ignored the correct data that was right in front of them (Klein, 1988).
- People rationalize information to fitting with the desired or expected outcome.

Meeting the user's goals and information needs means providing information which can be understood within the situation's context. The design problems arise because taking actions to achieve a goals requires comprehension of information which requires understanding both the information and the relationships inherent in that information. Influencing this construction of relationships is coherence of information as a positive factor and cognitive overload as a negative factor (Thuring et al., 1995). Interestingly, as the information flow increases in complexity, contrary to intuition, cognitive overload prevents increasing the complexity of the mental integration strategy.

Looking at informational system design as a cognitive problem conflicts with a design viewpoint that system analysis should focus on explicit user behavior, not user thought processes. Yet, with complex situations it is imperative a shift in view occurs. Norman (1986) concisely justifies the need for considering psychology by explaining user needs in terms of psychological goals and physical systems:

> There is a discrepancy between the person's *psychologically* expressed goals and the *physical* controls and variables of the task. The person starts with goals and intentions. These are *psychological* variables. They exist in the mind of the person and they relate directly to the needs and concerns of the person. However, the task is to be performed on a *physical* system, with physical mechanisms to be manipulated, resulting in changes to the physical variables and system state. Thus, the person must interpret the physical variables into terms relevant to the psychological goals and must translate the psychological intentions into physical actions upon the mechanisms. This means that there must be a stage of interpretation that relates physical and psychological variables, as well as functions that relate the manipulation of the physical variables to the resulting change in physical state. (p. 33)

If the cognitive factors (psychological variables in Norman's terms) play a part in how a person expresses and achieves their goals, then obviously those factors must be allowed for. Beyond that, any cognitive factors that operate as constraints must be allowed for as they too affect how a person expresses a goal.

Both an person's knowledge about the topic and reading ability affect how they shape goals and need information presented. With a dynamic information system, the text can be modified to accommodate these requirements. The content should change on a person-by-person basis to best fit their needs. Unfortunately for people currently designing complex systems, how to actually create information which cleanly supports such modification is still an open research area.

### *Environment constraints*

Environmental constraints are the physical and social factors that exist within the complex situation and impose barriers to defining and achieving goals or information needs. Perhaps the room is too small or hot, or an loud open work space that requires noise protection. And then of course, there is always a pointy haired boss with unreasonable demands on cost and time of implementation, or limiting choices to fit his expectations.

Environmental constraints are tempting to ignore since the "real information" about the situation does not include environmental constraints. Often, early in a project everyone acknowledges that other elements outside of the system (or available databases) come into play and are important. In design meetings on past projects, I've heard the acknowledgment range from refusal it consider it: "that's outside our scope, don't worry about it" to admitting it exists, but avoiding the issue: "good point, but we don't control that, the user will just have to take care of that themselves." Unfortunately, often even positive acknowledgement to consider the issues further become pushed aside as the designer gets involved in the details or as deadlines loom. Buried beneath project details, unless the original high-level design clearly accounts for environmental constraints, the designers simply has a hard time remembering to include those elements.

Although easy to ignore, the factors just mentioned most definitely influence how a person sets goals and goes about achieving them in a complex situation. Whereas many analysis methodologies might miss or discount these factors, and in a simple situation they can be safely discounted, in a complex situation they are just as important to consider as directly relevant goals and information needs. It is often these environmental factors that provide the nonlinearity and drive part of the dynamic information aspects.

### Technology constraints

Technology provides many constraints on the design of an information system and can often interfere with achieving a goal. A major technology problem is how it influences getting the data from the situation to the user. The data needs to be collected and stored before the user can access it with the information system. How much data is collected and how often are too major issues. High costs of real-time data collection prevent many systems from obtaining timely information. For the static data within a system, having people generate it can also be prohibitive. Finally, the expense of developing applications that work with the data can severely limit how much information is provided.

Improper technology can break the system, but it can never make it. One pitfall which forms it own constraint is when the design gets swept up in the latest technology. A system goal becomes of using the new or niftiest tool and can fail to adequately must focus on how the information and presentation will affect human behavior (Carliner, 2001). As Heath and Luff (1996) state, "it has been recognized that many innovative and advanced applications fail not so much as a result of technological inadequacy, but rather as a consequence of the systems 'insensitivity' to the ways in which individual ordinarily interact and collaborate in the workplace" (p. 98).

Designing a system which effectively provides information to users when and how they want it requires leaving the technology hype behind and focusing on the real information needs and user goals. Making a system usable means understanding the user's needs and providing for them. Technology can wrap, package, and ship the information, but it doesn't provide the information in the first place because it can't define what information it should be wrapping and shipping. That is the job of the information designer as part of the initial design and ongoing maintenance. In the end, effectively addressing the user's goals does not depend on how tools used to present the information, but on providing an adequate flow of information to the user in a form that makes sense in the situational context.

Adaptive hypertext and single sourcing coupled with a content management system are the current technology of choice for design of systems for complex situations. By providing an automated way of defining user profiles and populating templates with reusable information elements, they provide the technology required to present the information. However, while there is a great amount of research into the technology issues, there seems to be a dearth of research into how to create the content for these systems.

## Emergent properties of achieving goals

With the dynamic and multidimensional aspects of a complex situation, the user's goals constantly adjust to fit the current view of the situation. Thus, how the user understands the situation emerges as a result of interacting with it. In other words, the goals become an emergent property of the situation.

In chapter 2 as the characteristics of a complex situation were discussed, the issue arose of how users answers their current open-ended question by seeking sufficient information. Information needs emerges as a part of that interaction; finding one piece of information may lead to needing another or may render a third piece of information irrelevant. As a emergent property, how the goal is actually framed and what is considered acceptable with respect to achieving that goals varies based on the dynamics of the situation. What one user considers acceptable for declaring a goal achieved does not necessarily mean all users would consider that level acceptable. Other users might either feel the goal is not achieved or has been overachieved.

A user enters into a complex situation with a set of basic goals. By collecting information and through the feedback provided by the situation, the goals are modified and refined until the situation has changed to an acceptable state. Goals are not static; the user never has a complete set of goals when first encountering the situation. Instead, the user has ill-formed sub-goals based on an overarching goal. Through interacting with the situation and gaining a better understanding of it, the sub-goals emerge as firmly formed entities.

To aid in the emergence of the sub-goals, a designer must attend to the emerging structure of the goals and information needs at all times. These goals and information needs are in no way fixed and a design build on assumptions of fixed goals sits on a fast track to failure. The emergent properties arise at all levels. At the microscopic level of individual textual elements, at the level of information about discrete tasks, at the level of interactions and relationships between tasks, and eventually to the level at which the system interacts within the user environment, the new goals are emerging and existing goals are being modified as the user gains a clearer understanding of the situation.

## Wizards

Wizards have become a common technical solution of most major software programs to handle complicated situations. By activating the proper wizard, you can quickly accomplish a basic task. For designing a chart, the wizard can get you close enough to modify it (an approach which interestingly assumes you could design the chart without the wizard), or it can help get a simple database designed. As such, they are a very successful technique and should be used where appropriate, but they do not apply to complex situations. Refer back to the car buying example at the end of chapter 1 for an example of how a wizard and a complex situation approach differ.

The fundamental problem is that a wizard depends on having a fully defined decision tree, a characteristic of simple and complicated situations. A wizard is a linear procedure wrapped up into a special interface which rigidly follows the decision tree; a procedure which fails when something slightly different is required. A wizard does not allow more than trivial variations in a procedure, a technology choice usually made to keep the implementation simple. Yet in complex situations, the variations are the basic aspects that distinguishes the situation as complex versus simple or complicated.

In complex situations, people need something more than the automation of a wizard. They don't repeatedly solve the same problem, but variations. With variations, the wizard fails to fully address their needs and users face the choice of completing the wizard and hopefully fixing results, or simply not using the wizard.

**Example 1** Consider the use of a home medical diagnosis program and the choice is to say if your stomach pain is a sharp or a dull ache. But your stomach doesn't hurt at all. Should you enter a dull ache? Can you trust anything that comes after this questions, including its diagnosis of your problem (a rare tropical disease contracted from the bite of a worm that only lives in the Amazon, and you've never been to the Amazon).

**Example 2** I had a new computer from one of the major PC vendors which refused to turn off. It would go through the entire shutdown routine, seem to actually turn off, and then reboot. When I tried to used the wizard on the vendor's web site, after picking the model, the troubleshooting wizard asked for which piece of hardware was giving the problem (modem, hard drive, floppy, video card, etc). At this point, I had to stop using the wizard since it did not provide anything relevant to my information needs. Luckily, the technical support person immediately knew the real problem was a BIOS setting But BIOS settings were not included within this particular wizard's decision tree. This points out two problems with trying to use wizards for complex situation: Troubleshooting is a complicated or complex situation and, thus, it's hard to define the entire decision tree, and without a complete decision tree, the user is left hanging when the questions make no sense in the context of their current situation.

The ability of the system to change the amount of detail and vary how it presents information relationships provides direct support for emergent properties. As the user's goals change, the system can change what it presents to match those goals.

## Ways to meet user goals

This section discusses ways to connect the situation with the user's goals. It covers both the difficulties in meeting their goals and methods of understanding the goal requirements.

### *Inability to clearly define complex situation goals*

The biggest problem with meeting the user goals for complex situations is that the goals cannot be clearly defined. The previous section discussed how the goals were an emergent property, an aspect which prevents collecting them in advance. The complexity inherent in a complex situation prevents a rigorous analytical approach and the anomalies encountered in considering solutions prevent defining a single path.

Instead, for complex situations, the design analysis must provide for a deeper understanding of users, users' goals, and users' work processes than are common in current system designs (Carroll, 1995c). With the closed-ended questions asked in simple situations, the user goals can often be considered either a learning problem or a set of step-by-step procedures. Once the user knows what they need, they can look for it and expect the system to provide a clear answer. On the other hand, with the open-ended questions of complex situations, the ability to clearly define the user goals becomes difficult. There are many variations which enter into the picture and complicate the designers job of presenting what the user needs. Effectively providing information to allow users to achieve their goal does not require an annoying or dictatorial interface that forces them down a single path toward the "right answer," but an understanding of their goals and the information and flexibility needed to reach them (Quesenbery, 2003).

### *Capture goals for complex situations*

Defining the user goals and information needs for a complex situation is one of the hardest parts of designing a system that support complex situations. The dynamic information aspects and emergent properties of goals make it very difficult to clearly define them. A task which is straightforward in simple situations becomes fraught with difficulties in complex situations.

Rather than forcing the user into addressing the situation the computer's way, the designer must "start with the user's preferred way of solving the problem and to examine its strengths and weaknesses carefully" (Cohen, 1993, p. 266). This quote by Cohen reveals underlying logic behind the user-centered design movement as it represents the shift from focusing on user interface technology to focusing on the users.

A major problem of understanding a situation and defining the user goals is understanding the underlying problems of the situation as perceived by the user (Wickens, 1992). Thus, the users' main need is information which supports gaining a clear understanding of the situation and the possible solutions. The designer must gain an understanding of what information is relevant, how the information is obtained, and how it relates to other information. The knowledge about the relationships between pieces of information must be captured early in the process and allowed to drive part of the design process. An interesting aspect of analysis for complex systems occurs because they contain situations which cannot be specifically foreseen but for which the information requirements can be generally anticipated (Hutchins, 1995).

## Goals in hospital information systems.

Vicente (1999a, p. 38) reported on observations made by Hovde, who made several observations about differential success of the various computer-based information systems that installed in the hospitals he studied. He studied the productivity of the various software applications various hospitals had purchased. In general, they purchased off-the-shelf systems from a vendor and had problems. Because they were cumbersome to use, nurses and doctors either bypassed the system or used them grudgingly because management forced their use.

One hospital stood out because the system was used willingly and seemed to work well with high user satisfaction. The difference was that rather than buying an off-the-shelf system, it had been custom designed as a joint effort of the programmer and the hospital staff.

> An interesting phenomenon occurred during the programming phase. Being trained in logical thinking, the programmer reflected on the functional specifications as he was implementing them and came up with "more rational" ways of defining the functionality of the information system and the structure of the work flow. When he voiced his "improvements" to the future users, his ideas were almost always dismissed. Although the changes seemed logical to the programmer, they were not based on a thorough understanding of the hospital's work demands. However, this was only obvious to the workers because they were the ones who had a deep appreciation for the factors governing the context in which the information system was to be used. The programmer did as he was told, and implemented the functional specifications developed by the workers, rather than changing them to fit his own criteria. As mentioned earlier, the result was an information system that was used and that seemed to achieve its goals in an effective manner. (Vicente, 1999a, p. 38)

In this example, the users had a different and much deeper view of how the system must operate than the programmer did. For systems to support complex situations, the designer must work closely with the users to begin to gain that deep knowledge and to ensure the designer view does not add a unwanted "improvements."

To overcomes the problems identified in the preceding paragraph, both Endsley (1995) and Marchionini (1995) claim the analysis must identify the major goals and the sub-goals necessary to meet the goals. As part of defining the sub-goals, each of the major goals needs to be identified along with the information needed to support clearly understanding the goal's relationship to the current situation. Besides just defining the information needs (all too often the end point of task analysis), how that information is integrated or combined in the process of understanding the situation needs to be addressed.

The metadata required to connect the goals, information, and relationships directly arises from this analysis. Only through the metadata attached to the various information elements can the system dynamically assemble the proper information for a user.

The next few paragraphs examine each of these major steps. This section focuses on defining goals and relating those goals to information needs. A detailed discussion of defining information needs is deferred to the next chapter.

**Identify the major goals and the sub-goals necessary to meet the goals.** The goals in a user's mind at each stage of understanding the situation and achieving any overarching goal must be collected. As a part of defining the sub-goals, how those sub-goals might be modified as the situation changes must also be considered. In addition, analyzing the methods of accomplishing goals in an complex situation requires identifying the goals that arise in coping with unfamiliar and unanticipated events in a flexible manner (Hutchins, 1995).

Create a hierarchical diagram that contains the goals and sub-goals (see the appendix for more details). It should also include the branches that handle error conditions. As there is no single path, sub-goals could conceivably appear on multiple (side-by-side) branches on the diagram, with each branch representing an independent route to achieving the goal.

**Define the information needs to meet the goals.** Each goal has information which identifies:

- The relevance of the goal to the situation. Is this goal even relevant to the current situation? Because of a unique aspect of the situation, a goal may not apply. Also, it may not be currently relevant, but will be as the situation changes.
- The exact nature of how to shape the goal. Goals are an emergent property of the situation. This information helps to refine and shape how the goal fits within the context of the current situation. Because each complex situation is slightly different, each time the situation is encountered, the goals are shaped slightly differently.
- If the situation is changing properly with respect to achieving the goal. How should the complex situations change as it transforms from one state to another. More than just the end points are required, but also an understanding of the information which shows the transition is proceeding properly.
- If the goal has been accomplished. When has the situation changed to a state that achieves the goal?

As each goal and sub-goals are identified, all four types of information need to be defined for each one. An important aspect of information needs within complex situations is that not all users use all the information. Information needs must be collected and defined across a cross-section of users and not just one or two experts. When the users works with the system, the information manipulation ability will allow them to refine the presentation to only show the information they need.

**Define the information needed to clearly understand the goal's relationship to the current situation.** Goals do not exist in isolation, they are connected with other goals. As a situation changes, the interconnections change in importance. How those connections are formed and broken and what factors influence the forming and breaking of connections need to be defined. Also, how will changes in the situation caused by achieving one goal affect achieving another goal. A complex web of relationships quickly form that can impede achieving all the goals unless the relationships are uncovered and allowed for in the design.

The goal's relationship to the situation also brings in the information needed (not necessarily wanted) to understand how the goal relates to the situation. People tend to ignore information they do not want to see or if they do not understand how it applies to the current goal. Providing the needed information means defining both what information is needed and its relative importance to ensure all information is given proper salience.

**Define how that information is integrated or combined in the process of understanding the situation.** In what order do people form goals and how do they use the available information to achieve the goal? The purpose here is not to define what information is needed, but to define how people use the information they need. This step allows the design to provide information in a coherent manner relevant to the situation.

One way to determine this is to identify the current heuristics and methods used to handle similar situations under the current system. These methods have evolved to fit the current social and environmental constraints; until the reasons for their existence are understood, they cannot be ignored. However, just because they exist under the current way of doing things is no reason to carry them forward. But their reason for existence needs to be identified to make a conscious, informed decision about their relevance to the new design (Hutchins, 1995).

## Conclusion

As the design analysis has shifted from simply listing needs to including definition of interrelationships, the data collection problem moved from simple data gathering to fitting into the real-world model and people's understanding of relationships within that model.

> In our experience, superior designs result by focusing initially on the nature and structure of the work and then later adapting the user interface to operating conditions and constraints as the design details emerge. (Constantine & Lockwood, 1999, p. 36)

Designing a system that helps the user achieve goals in complex situation is a tall order. A design problem that is itself highly complex. Yet, information system must start to provide this type of support if they are going to continue to help improve productivity and meet user expectations. An effective design analysis of the user goals for a situation provides a designer the foundation for providing readers with highly integrated information which reduces the user's cognitive workload and allows easy understanding of the situation. After defining goals, the type of questions people may ask and the type of answers they expect from the system can be defined and developed as a fundamental part of the system design.

When faced with a complex situation, the user develops goals that address how to understand the situation. Embedded within the system's logic must be the basic structure of the goals and information needs found in the analysis. The user interacts with the system to manipulate this information to refine the view of situation and maintain a clear picture as it moves between states. Normally, gaining an adequate understanding involves relating situation information across multiple goals. Thus, providing a design that meets the overarching user goal of understanding the situation, requires a design that matches user goals to information needs. As the user interacts with the situation, the system must capture the dynamic information changes within the situation and relay it back to user in an effective presentation. Here an effective presentation is defined in terms of supporting user goals, helping the user refine or develop goals, and having salient information at the proper detail level. A detail level which can be easily modified as needed.

The people viewing the information have a goal of locating the relevant information, mentally forming the relationships within the information, relating it to their real-world situations, and, if required, using it to perform a useful or correct action. As always, the emphasis is on the relationships between goals and information, not the information itself. Many system's failures arise not from the failure to provide data, but the failure to anticipate the user's real-world goals and information needs. Even with high information content, because of poor design, the information cannot be effectively transmitted and becomes both hard to find and hard to process (Bowie, 1996). Being able to look up information and then compare or compile that information forms a fundamental design aspect when addressing complex situations. Even user tests reveal this bias when they focus on answering questions with simple answers, such as "What was total sales in the eastern region in March?" Yet, people rarely actually need this single piece of information. Instead, they want to understand what factors contributed to the eastern region sales in March, how those sales compare to previous months, and what can be done to increase them.

As Norman (1988) wrote: "An enjoyable and effective user experience does not come about accidentally; it requires considerable focus and attention to the needs, abilities, and thought processes of the users" (p. 40). Providing an enjoyable and effective user experience with respect to complex information needs requires uncovering and defining the user's goals at multiple levels, the information they need to address those goals, and the interrelationships inherent within the information that makes it relevant to the current situation. These are issues that lead us into Chapter 4 "Information needs."

## Long example—Business report analysis

This example considers the analyzing of business reports, a complex situation with constantly varying user goals. Business report analysis qualifies as complex because it does not lend itself to being performed using a defined set of tasks nor can those tasks be performed in a fixed order. No matter how detailed the analysis, a step-by-step description for understanding the entire production environment or solving sales problems cannot be obtained. Instead, the initial system design analysis needs to reveal the general information needs and the potential sub-goals the manager may address as they proceed to gaining an understanding. A manager may use a set of guidelines (heuristics) for analyzing reports and spotting problem, but these must be adjusted for each individual situation or problem. Each new piece of data the manager uncovers affects the path taken and the eventual outcome.

Supporting the analysis of these reports means getting into managers heads and figuring out how they perform the analysis. Then, by supplying dynamic online information, the report can be customized to their current goals. The problem with performing this customization is understanding the managers, not getting information. As described in chapter 1 of this book:

> The information the analysts needs exists. The data used to create the standard report is in a or more relational databases, and textual information such service contracts and memos are in the content management system. What is missing is the ability to connect them into an integrated presentation that assists the user.

Think of these basic design problems when supporting report analysis (the list can easily be extended):

- Each person works in a different way when analyzing reports.
- People probably cannot clearly articulate how they interpret reports; they just do it. And, depending on time constraints, some months they might look at them more than others.
- The information of concern to people differ depending on their management style.
- The researching and handling of problem situations is quite different from checking that everything is ok.

Designing an information system that addresses these problems is a highly complex problem. Paper-based reports could never accomplish it, but a system with dynamic information generation might have a chance if the early analysis gained a clear understanding of how the information was analyzed.

The method described here can help with the routine analysis and provide an indication of a problem. It will not necessarily be able to identify the source of the problem or how to fix it. Unfortunately, problem resolution is a management skill that cannot be automated. In other words, the designer should focus on clear information presentation to assist the manager in effectively evaluating the reports and understanding the business situation. This is not a method for automating business management.

## *Addressing the goals of business report analysis*

Work with a broad base of managers to determine what they consider the major goals to be, goals that must be better articulated than "ensure my department is running ok." What is most important: budget, productivity, employee retention or satisfaction? There will be along collection of goals, with each only applying to some managers. However, as they will all be using the resulting system, it must support each person.

### Identify the major goals and the sub-goals necessary to meet the goals

Working with scenarios that make the users focus on specific techniques can help draw out their goals and the order in which they set goals. It can also help determine what goals they consider irrelevant and why. Also, methods such as Critical Incident Technique (Emmus, 1999) can help determine goals when the overall situation starts to depart from routine.

- What goals are set in looking at each major area (budget, productivity, sales)?
- What questions need to be answered to understand if a problem exists?
- What questions need to be answered to ensure the situation is normal?

### Define the information needs to meet the goals

Once the goals are set for each scenario, work through and define what information is required to meet the goal.

- What information shows the situation is normal?
- What information is most important?
- What information is necessary and which is only nice to have?
- What information is only looked at if other information is outside a specific range? What range?
- What information can lead an inexperienced person to think a problem is occurring? What can defuse that misunderstanding?
- What happens in high- or low-end situations? For example, Christmas rush, months of low production, or during various phases of economic downturns.
- What are the information values that are too high or low to be realistic and probably indicate a problem with the data?
- What information would change first to help reveal a problem early?

### Define the information needed to clearly understand the goal's relationship to the current situation

For any situation, there can be a large number of possible goals. Yet an effective, experienced person quickly eliminates them, normally so quickly that Klein (1999) found they would say they didn't consider them. Define what makes the goals the user thought were important relevant to the situation and what makes other goals irrelevant. Goals that at first glance seem to be relevant or might be relevant to a related situation should be specifically addressed. What information provides clues to distinguishing and differentiating them?

- How are the goals related? In what order are they addressed?
- What orders for addressing goals should not be used because they can cause incorrect but logical conclusions?
- What information is often ignored or missed but helps to clarify the situation?
- What information should always track together? In other words, both values should vary together in a predictable pattern.
- What supporting information, not in the system, can be used to verify information or help understand the situation?
- What information would show that a potential problem is not really an early problem but normal fluctuation?

### Define how the information is integrated or mentally combined

Watch people work within the situation and notice how they actually handle the information and collect other information from the situation.

- What is the order the information is looked at to verify the situation is normal?
- What is the order the information is looked at to analyze problem situations?
- How should the information be presented (graphs, text)?
- What level of detail is required?
- How does the level of detail change and what motivates the change?
- How will the information be manipulated?

### *Human-computer interaction*

The system is incapable of performing the report analysis; instead, it must act as an expert consultant by providing guidance and information in a form needed by the business analyst when they are needed.

Some of the items the system should provide easy access to include:

- Options of things to look at (low or high sales) and how they should be analyzed. This can range from a simple checklist for an experienced analyst to more detailed instructions for a new analyst.
- List of any special executive management or customer concerns that should be discussed this month but are not part of the routine report analysis.
- List of previously identified problems or aspects that merit closer examination to ensure they are improving.

These items need to be more than just comments like "Check eastern sales." Instead, the pertinent values for eastern sales, relevant modifications (such as promotion information), historical information, and any other information deemed relevant for analyzing Eastern sales should be presented. The actual design would probably split into two different type of analysis. The first is routine analysis. Checking Eastern sales is one example. Each report period the analyst would expect to perform that task and would generally perform it in the same way. A dynamic information system can be very effective at addressing this type of analysis. The second type is the ad hoc analysis that would be done to explore the underlying causes of a problem. Here, the system can supply pointers to ensure that potential paths are not missed (they would be noticed by using methods such as critical incident analysis).

# Information Needs 4

*Those who cannot tell what they desire or expect, still sigh and struggle with indefinite thoughts and vast wishes.*
*—Ralph Waldo Emerson*

*Confusion and clutter are failures of design, not attributes of information. And so the point is to find design strategies that reveal detail and complexity—rather than to fault the data for an excess of complication. Or, worse, to fault viewers for a lack of understanding.*
*—Edward R. Tufte*

Chapter 3 discussed user goals and how a user's desire to achieve them affected a situation. Throughout that discussion, the information that the user needed to understand and track the goals was just assumed to be available. In this chapter, we look at how to define what information is needed and how to present it.
In chapter 1, information needs were defined as:

> The information required by the user to achieve a goal. These information needs often require both information to initially understand the situation and to access that the goal was achieved. A major aspect of good design is ensuring that the information is provided in an integrated format that matches the information needs required by the user goals. Although a perfect system always contains the information needed to meet a goal, in reality, the design must compensate for incomplete information.

This chapter expands on that definition and considers the details of how information connects with the goals set to understand a complex situation. As part of that consideration I:

- Examine how people use information to achieve goals and how different information presentations can affect achieving a goal.
- Formalize the definitions of data, information, and knowledge.
- Explore how understanding of information comes not from the information itself, but from the relationships found within information.

## Introduction

With information needs, we finally focus on the content which users requires to address their goals. Interestingly and, perhaps, unfortunately, the content often gets short-changed in many design discussions. The problem is that content is normally assumed to already exist, can be used as is, and, thus is outside the scope of consideration. Yes, the content is situation specific, but it still will never just appear out of nowhere in a fully developed form. Or it will not already exist in a database, ready for the system to draw on. Also, as a person interacts with a situation, the information they want for any particular goal changes as they get a better grasp of the goal and situation. This chapter works to provide a means of connecting user goals with user information needs, so that the designer and writer can build a stronger, more effective informational system.

As a simple example, of information needs, try to use the information in Figure 4.1 to get a grasp on the current produce production situation. The table does contain the information a person needs, but the arrangement complicates the analysis. Because the table is ordered by apricots, not by production year, it can be hard to get a clear picture of any other produce or even how apricot production changed year by year. As an example table, Figure 4.1 only contains a few rows: think of the problems if it contained 50 or 100 rows or 20-30 columns. Although this problem could be partially solved by providing a method of letting the user sort by a selected column, in most complex situations, the information is not as clear cut nor does it lend itself to a single table. Often the user does not just want lemon production in 1983, but needs to be able to analyze it with respect to both plums and grapefruit production across a five-year interval. Complex information contains multidimensional structures and complex relationships inherent in the information which prevent straightforward solution.

Tree Fruits Per Capita Consumption, 1976-77 to Date

| Season | Apples | Oranges | Lemons | Apricots | Grapefuit | Plums |
|---|---|---|---|---|---|---|
| 1978-79 | 5.02 | 8.11 | 1.28 | 1.08 | 1.51 | 0.23 |
| 1982-83 | 8.01 | 12.84 | 0.97 | 0.88 | 4.75 | 6.32 |
| 1981-82 | 9.74 | 5.55 | 1.89 | 0.54 | 3.10 | 2.56 |
| 1977-78 | 7.34 | 8.76 | 1.14 | 0.52 | 3.42 | 3.41 |
| 1980-81 | 9.22 | 5.66 | 2.12 | 0.50 | 4.97 | 4.10 |
| 1979-80 | 3.51 | 14.68 | 2.22 | 0.29 | 2.84 | 3.50 |
| 1976-77 | 7.11 | 6.16 | 0.60 | 0.24 | 4.04 | 4.76 |
| 1983-84 | 3.39 | 7.64 | 1.71 | 0.23 | 3.81 | 6.19 |

FIG. 4.1. Poorly presented produce information. Although neither the dates nor the produce are ordered, the table can still be used, just not very efficiently. This table is sorted by apricots. Would you want to use to use it to get an understanding of grapefruit production over the last five years?

The problem of addressing information needs extends well beyond having the information available and even having it well arranged. Once the problem becomes complex, people find it hard to figure out what information they need. One study found that approximately half of the participants (i.e., college students) failed to extract the proper information for ill-defined problems when the relevant graphs and illustrations were presented to them (Guthrie, Weber, & Kimmerly, 1993 as cited in van der Meij, Blijleve, & Jensen, 2003). Consider how much more difficult this can be when a user either does not know or is not sure the information exists within the information about the situation. The user will never search for it or, if she does search for it, will get frustrated and question all the information quality when it can't be found. Yet, the designer's or writer's job is to ensure the user knows the information exists, does extract the proper information, and understands, in the end, the situation to which it applies.

A person working in a complex situation needs more than just the data from Figure 4.1, they will also need some level of supporting text. When the need for this text is considered, their information needs move even further beyond the simplistic solution of providing methods of sorting the table. How much supporting text depends on the specific goals and the knowledge level of the user. Two users looking at lemon production over the past five years can be doing so for very different reasons. With dynamic information presentation, both the amount of table information and supporting text can be adjusted to fit the user's needs.

## Information needs to achieve goals

Goals could be viewed as the lofty plans and the road map to follow, but it's the information that makes up the roads connecting the situation to the goal. Only through the information coming out of a situation can a user understand it and make informed decisions about how to proceeded. Any complex situation contains an over abundance of data. Data that needs to be collected by an information system and feed back to the user on demand. As such, with complex situations, the user needs clearly structured information that helps reveal solutions to the open-ended questions and provide connections across multiple-task procedures. Collette (1991) succinctly states this goal: "To help capture, organize and deliver the knowledge of how to act, in a timely and effective way, so that the appropriate action will take place again and again" (ET-45). The system should enhance the users' ability to integrate and manipulate information into inventive solutions to their unique problem. Achieving this goal requires knowing what information is required, how to manipulate the information to extract the required knowledge from it, and how to construct mental models of the situation which can be used to handle unanticipated problems (Brown, 1986). This approach in turn means taking into account the dynamic nature of the situation's relationships and connects to how the situation's history affects its future.

Designing for easy manipulation and assimilation of information requires minimizing cognitive effort of activities not directly tied to the information. "Every effort additional to reading reduces the mental resources available for comprehension. With respect to hyper-documents, such efforts primarily concern orientation, navigation, and user-interface adjustment" (Thuring, Hannemann, & Haake, 1995, p. 59). Because any complex system is dynamic and constantly changing (Cilliers, 1998, p. 40), the system requires a higher level of control to effectively handle the manipulation of the information. The information needs in complex situations are not static; they change and emerge as the person gains a better understanding of the situation.

Information comes from all parts of the situation and does not just reach the user via the informational system. Issues such as office politics and previous knowledge play a major part in understanding a situation, but will never be in a form to be captured and fed to the user (Figure 4.2). Expert behavior with respect to understanding a situation arises from having an understanding of these relationships; an understanding which is often developed to the level of "intuitive feel" for how to respond. This can include actions such as better setting of subgoals, but can also be knowing which information to pay attention to and how to interpret incomplete factors.

The user does not want to just have information; the user wants to evaluate or analyze information to find solutions with respect to a specific situation. The question becomes one of how to ensure "whether the information will be logical to the user in the context of the use of the information" (Duffy, 1995, p. 269). In helping to situate information properly for the user, Marchionini (1995) lists three dimensions for characterizing information: "specificity, quantity, and timeliness" (p. 37).

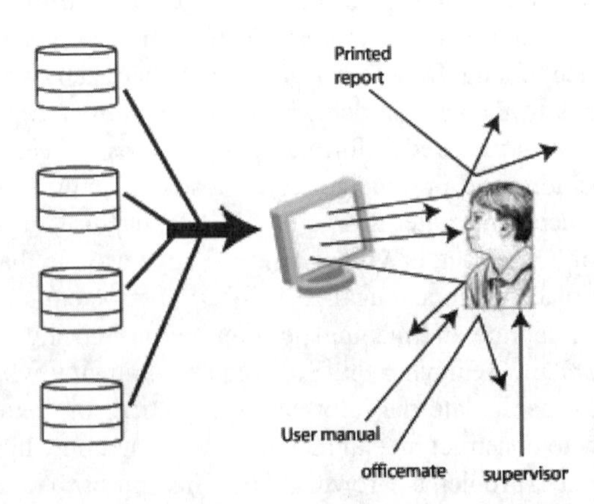


FIG. 4.2. Information flows from more sources that just the database. Ignoring these other factors can lead to a system of questionable usefulness. How many of the arrow bounce off the user depends on both experience and how clearly the information is presented.

**Specificity** Specificity considers how directly applicable the information is to the current situation and the current goals. In the ideal presentation, the reader receives only the required information in the proper order. Any extraneous or irrelevant data decreases the specificity value of the information, as does having to perform extensive manipulation to place it into an understandable format. Irrelevant information may also cause the person to make the wrong choice (Hsee, 1995).

In complex situations, defining the directly applicable information becomes one of the hardest parts of the entire design problem. Because a situation has a history, the historical effects strongly influences specificity. The history also interacts with the multidimensional aspects to determine the relative importance rankings of various pieces of information.

What users need from the information is to gain the knowledge required to understand the situation in its current context and, normally, make some type of decision about it.

**Quantity** Quantity is how much information the reader gets. Too much leads to information overload but not enough prevents the reader from forming a clear mental image of the situation. The data drill-down implemented in most online data systems provides a simple method of controlling the presentation quantity but has a difficult time showing relationships.

In essentially all complex situations, the design problem is not one of lacking information, but one of overabundance. The appropriate quantity of information must be combined with the specificity requirements to provide the proper quantity to both prevent information overload and describe the situation.

For any quantity of information to enhance productivity, it must extend into the users' environment and provide the bridge that gives access to the data's full potential. This does not necessarily mean full data access (obviously factors such as business data security prohibits such an idea), but it must provide the integrated data presentation to allow readers to maximize its benefits.

**Timeliness** Timeliness is having relevant data soon enough for it to be useful, and only having it while it is useful. Information has a time period when it is useful for the user with respect to the situation. Problems can occur when information is presented too soon for the user to make sense of it or too late for it to be useful.

Because information tends to be dynamic, only the most current information should be displayed. Old or out of date information is worse than useless because it can be misleading. The frequency of update needs to be found during the design analysis as all major pieces of information about the situation changes at different rates.

The timeliness of information combines with its specificity to provide high quality. Information must be managed carefully to get rid of what is irrelevant and out-of-date. Plus, the user need helps that help with better search, and to be able to judge the quality faster (McGovern, "Information Technology", 2002).

## Shuttle puzzle example

The amount of information a person needs varies by the situation and previous knowledge. In Figure 4.3-A, the three pieces are insufficient to provide a clear picture. A rocket launch might be obvious, but could you say for sure it was the space shuttle? Yet, in Figure 4.3-B, there is no doubt that the image a space shuttle launch even if all the pieces are still not present. There is nothing special about the missing pieces, Figures 4.3-C and 4.3-D all have the same number missing as Figure 4.3-B. In each picture different pieces are missing, yet in each the overall image is understandable. Quick question: if you only saw Figure 4.3-D, would you say the shuttle has an exhaust flame? How do you know?

As compared to Figure 4.3-A, the image in Figure 4.4 is missing the same pieces, but the arrangement of the provided pieces is random. Could you tell this was a space shuttle launch? Unless the information provided to the user fits within their expectation and has an arrangement that makes sense, even with sufficient information, the overall pattern cannot be discerned.

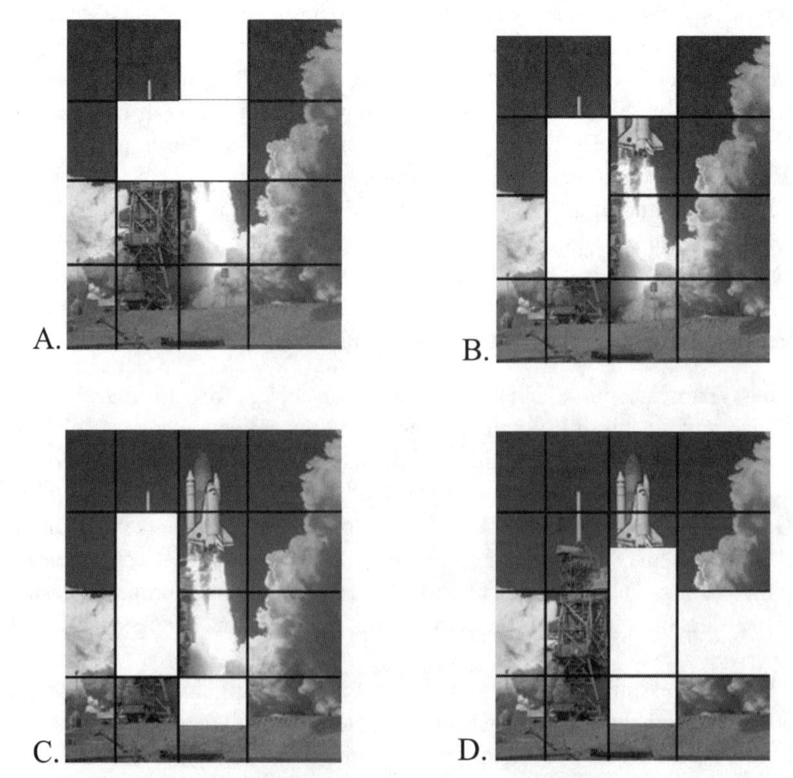


FIG. 4.3. Effect of different pieces of missing information

Figures 4.3-A and 4.4 contain the same information, but chances are good that a person receiving the information in Figure 4.4 would never understand the situation. Having a person who had never seen a shuttle launch to pick out these pieces from a collection of other pieces would be almost impossible. The relationship information that connects the pieces is missing. Post-event analysis of information presented in a way comparable to Figure 4.4 (normally after a major wrong decision occurs) would reveal that the user had all the necessary information. After carefully filtering and arranging it, the overall situational context becomes obvious. Yet to the original user, the information was incoherent or invisible.

On *Wheel of Fortune* or *Concentration* the game consists of exposing information provided piecemeal until the answer is known. The exact letters or images that will be exposed before the contestant guesses the puzzle is unknown. Watching the game, it can be amazing either how people guess the puzzle with almost nothing exposed or essentially everything is exposed and they still can't guess it. Likewise, users takes information and assembles a mental image of the situation and finally, often rather abruptly, reach what they consider an understanding. In simple situations, it's easy to define and ensure the information gets presented. In complex situations, it becomes a much more difficult design task. The unknown factors inherent in not being able to define which information will be seen complicates the information design problem. The information provider needs to provide information elements in an order that allows reaching an understanding rapidly and correctly, but simply cannot work from a complete picture of what will be seen.

FIG. 4.4. Space shuttle launch broken up like a puzzle with the pieces scrabble. These pieces match the visible parts of Figure 4.3-A, but provide no coherent view of the image. If a person had never seen a shuttle launch, properly assembling these pieces or deciding if they were part of the same picture would be difficult tasks.

## Hierarchy of data, information, and knowledge

Up to this point, I've been using the words "data" and "information," but have not defined the terms. In normal day-to-day talking, we tend to use the words data and information interchangeably as does the computer trade press. However, this book works with much finer grained definitions which builds a hierarchy of the three terms. On the other hand, the distinction between data and information varies by situation with what is data in one instance may be information in another.

### *Data*

Raw numbers, facts, and figures are data. Alone, a collection of data means nothing. A table listing daily stock values contains data, the numbers carry nothing more than a value. Are the numbers higher or lower than last year, last month? What caused a major jump between two consecutive days? Are the number so inconsistent that they make no sense and must be checked (think of external understanding required to even know the numbers are inconsistent)? If the table only contains the values, it cannot help answer these questions.

Data provides a foundation developing into information, but it must be combined and integrated with other data before it becomes useful in a more than trivial sense.

### *Information*

Information is data with semantic association. It relates to the situation and contains the relationships that give it the semantic associations. An information system's most important role is present the information so users can perform meaningful tasks. An effective information system must support the changing information needs and not focus on only one (Elam & Mead, 1990; Shneiderman, 1992), especially since users constantly adjust their plans to reflect changing conditions and information (Jirotka & Goguen, 1994). In other words, information is data in context. The readers look at the available data and apply it to a specific context or situation. Within that specific situation, the data has relationships between different data elements that assist the reader in interpreting and understanding it. Most of those relationships arise out of the situation itself and only make sense within a limited range of contexts. Thus, to be fully useful, the person must be able to modify or adapt the information. Analyzing the situational context to uncover and define the data and information interrelationships make up the most important part of supporting complex situations.

Gaining this understanding often requires bringing in situational information that is not contained in the data. For example, a manager knows certain numbers should be trending down based on long-term corporate goals. The data may show an up or a down trend, but it requires the manager to know other facts outside the data to understand if the trend is good or bad. Is the trend down too fast or too slow? What factors are driving the trend? Supporting problem solving in these situations requires understanding the questions that may be asked and having the data available to answer them.

Good information designers provide readers the ability to transform data into information because as part of the design process the information designer gains an understanding of the potential situations, and makes the relationships and contributing factors easy to extract from the data.

### Knowledge

A person rarely wants information; what they want is knowledge that can be used to accomplish a goal. The information designer must shape data to transform it into information that the reader can easily transform into knowledge. The information technology trade press has thrown around numbers that estimated the losses to U.S. companies to be over $12 billion in 1999 because of problems with maintaining and understanding the information and gaining knowledge from it—a number that, unfortunately, is expected to increase dramatically each year.

Complicating the transformation from information to knowledge is the fact that knowledge exists in two different forms (explicit and tacit), only one of which lends itself to easy capture (Desouza, 2003). Yet, the other is vital to making sense of the information.

| | |
|---|---|
| Explicit knowledge | Explicit knowledge has a tangible value and can be recorded. This makes it easy to capture and present back to a reader. The sum of all the information displayed on a screen is the explicit knowledge. |
| Tacit knowledge | Tacit knowledge is understood and used but can't be recorded. Making sense of the values on a report and acting properly is tacit knowledge. Filling in the gaps of incomplete information depends on the tacit knowledge possessed by the reader and often is a major factor in determining skill levels. |

The users viewing the information have a goal of locating the relevant information, mentally forming the relationships within the information, and relating it to their real-world situations. When they accomplish that mental process, they have transformed the information into knowledge. Unfortunately, this final step from information to knowledge is one that can only be imperfectly supported. Knowledge is the useful stuff that's inside our heads. Knowledge let's us do things, like design new products, answer support questions, sell a product or service. Most of the process occurs within the mind and involves many subtle real-time factors. The training and ability to make these type of transformations are why senior executives and business analysts command high pay.

Large volumes of data and information are not sufficient. Unless it is properly organized, people cannot efficiently process large amounts of information, especially with the incomplete information available in a typical real-world context. The power to transform the information into knowledge depends not on the amount but on the clarity and number of connections linking the information together.

Knowledge and cognitive control are independent. It is possible to have the information required to be knowledgeable and perform poorly or inconsistently (Andriole & Adelman, 1995, p. 23) . The person might lack the ability to fully understand the information (needs more training or experience) or may be unable to effectively process it (other demands are tying up the necessary cognitive resources). Likewise, a person can have the knowledge and still make bad decisions or formulate poor or improper goals for the situation.

## Baking analogy

This baking example is one of clearest analogies I have found to show the difference between data and information, with the ingredients as data, the batter as information, and the finished biscuits as knowledge.

> A cook is supposed to have said, "the flour itself does not taste good, nor does the baking powder, nor the shortening, nor the other ingredients. However, when I mix them all together and put them in the oven, they come out just right for biscuits" (Henry, p. 91).

But the flour is messy and shortening makes everything greasy, so why not just work with the nice clean ingredients to make biscuits. Everyone knows not using flour and shortening means no biscuits, but yet, when it comes to manipulating information rather than physical objects (baking ingredients) variations on this type of logic keep people from using information they know they should consider. Just because nothing will prevent shortening from being greasy, the head cook must not avoid providing it to the baker. No design can prevent all information from being messy, but it still should make it easy to handle.

### *Information is multidimensional*

Complex situations are multidimensional which means the information within the situation is also multidimensional. Dimensions which must be considered as part of the analysis and design.

In chapter 2, we saw that user knowledge, detail level, and cognitive ability were three different dimensions of effective information which had to be considered. The different user groups of a system make up another dimension. The information needs and their goals vary between different groups. Added to this various dimensions are the different ways that user's dynamically view and reshape their view of the information as they interact with the situation.

O'Malley (1986) described the users' information need in terms of dynamic structures for which "*users* should be able to structure or restructure the information to suit their own unique purposes" (p. 396). The task of using the information entails exploring that information space and finding the relevant and discarding the irrelevant information until the appropriate amount of information has been obtained. In the course of exploring an information space, the reader needs to:

- **Integrated access complex information.** The multidimensional nature of information in complex situations requires that it be presented with some level of integration that clearly connects all the dimensions. A user often lacks the knowledge to be able to or choices not to perform extensive mental integration.
- **Adjust the presentation to fit the current goals and information needs.** Goals and information needs vary between people and with the same person at different times. The amount and style of the presentation (i.e., detail level, assumed knowledge) must be dynamically adjustable to fit the current needs.
- **View the problem from multiple view points.** The situation often contains various viewpoints. Cognitive tunnel vision can prevent a person from seeing an issue in more that one way and, consequentially, can hinder gaining a clear understanding.
- **Understand the relationships between information elements.** The important changes often only exists as measured by the change in the relationships, not as a change in the individual tasks. (Cilliers, 1998)
- **Receive updated information as values change within the situation.** A complex situation is rarely static. As information changes, it must be conveyed to the user and the user must be aware that information has changed.

Achieving all of these points requires a certain level of creativity on the user's part, but it's a creativity that must be supported by the information design to assist in the transformation of information into knowledge. A well-designed system allows the user to get on with mentally analyzing the information, while a poorly designed one bogs the user down in simply locating it. For example, one system provides an integrated text while the other provides a collection of short articles, each on one particular topic.

The design analysis must map out the multidimensional information space. A highly complex problem in its own right since this requires uncovering the applicable dimensions, the information relevant to each dimension, and the relationships that interconnect the information.

A standard practice for handling complexity of the situated context is to attempt to limit the problem-space. Rather than attempt to handle the entire dynamic process, the analysis focuses on single slices by either time or task. Much of the structuring of information attempts to predefine the information needs; thus, causing system breakdowns when users try to go beyond solutions envisioned by the esigner. The analysis methods that attempt to define a static space ignore the highly dynamic nature of real-world information spaces (Bist, 1996). With a focus on static spaces, most analysis discussions ignore how the user processes information and focuses on information is used rather than how it is used and applied to a situation (Warren, 1993). Woods and Roth (1988) criticized this simplification strategy as being too limiting. They describe the problem of "reducing the complexity of design or research questions by bounding the world to be considered merely displaces the complexity to the person in the operational world rather than providing a strategy to cope with the true complexity of the actual problem-solving context" (p. 417). In other words, the snapshot analysis is invalid because information needs and relationships have changed; this is often described by the common complaint "the requirements changed."

However, since complex situations occur within a dynamic, multidimensional environment, the design must support the dynamic, multidimensional nature of the complex situation, not a simplified static situation.

## Stock price: data or information

These were the prices for two stocks at about 3 PM today: Dell $28.54 HP $27.28.

Is this data or information? In most situations, just knowing the closing price would be data, because of various reasons:

- The person may not know what Dell or HP are.
- The person doesn't have any interest in these particular stocks and may have no interest in stock investments.
- Even if the person has Dell and HP stock, those values are data because they contain no past history. Only a select few, such as stock brokers and people interested in buying or selling soon, would know the recent stock prices. Are these prices higher or lower than yesterday and last week?

For this data to be information normally requires providing contextual situation with other data, such as:

- Company information such as their earnings per share.
- Previous history (are the values a rise or decline in share price).
- Any recent announcements that can be predicted to cause a short-term spike or drop in share price.

Even if I were closely watching the share price with the intention of calling my broker to place an order when they reached a certain value, these values are still data. It's all the past information which allows me to interpret these values that allows me to make that decision to call my broker or not. Of course, if the Dell and HP quotes were combined with other data elements, then they could give the user information.

When structuring information we must understand the user so that we are not providing meaningless data when we intend to provide information. For example, don't just provide today's closing values when the person needs past history. And don't provide the entire New York Stock Exchange closing prices when all the person wants is HP and Dell. (Requiring a search based on ticker symbol is imposing an extra layer of high cognitive load which could be avoided in most situations.)

## Information interrelationships

The failure of many information systems arises not from the failure to provide information but from the failure to anticipate the users' real-world goals and the information relationships required to achieve the goal. Yet, designers cannot be simply faulted for the lack of good design, we still lack clear methods of how to perform that design which McGovern (2000) describes be stating "in the civilizing of information we are still at the Searcher-Gatherer stage." Our search engines are crude devices limited, for the most part, to matching keywords and proprietary ranking algorithms. Our displays are still mostly text with a veneer of non-contextual graphics, although the visualization research shows promise of releasing us from this problem. Easy ways of tracking, manipulating, and refinding information on intranets and the Internet are sorely lacking.

A fundamental problem to overcome to progress beyond the searcher-gatherer stage is that people have a hard time integrating information and relating various data points to each other. But even more importantly, they have a hard time remembering or considering subtle cause and effect, goal-oriented relationships that exist between the information being viewed and other relevant information. Understanding the relationships provides a solid foundation for understanding the cause of and how to solve a problem (Casaday, 1991) and places the entire situation in context (Endsley, 1995). Yet, almost by definition, this is the presentation and manipulation requirements for addressing complex information needs. That is, helping people with complex issues and open-ended questions requires providing contextually relevant information that fits their goals and information needs, and makes easily apparent the interrelationships within that information.

### *Creation of relationships*

Earlier in this chapter, we looked space shuttle pictures. Now consider this next picture (Figure 4.5) that uses the same pieces as the early incomplete images. Obviously, these pieces are out of order. However, you only know they out of order because you already know what a space shuttle picture is suppose to look like. You have a mental image of the space shuttle and can form relationships between each of these pieces and that mental image. If this was information about an image or a situation which you didn't understand, then it would be hard to both know that the information was out of order and which information was missing. In other words, if a clear mental image was lacking it would be impossible to create the interrelationships between the pieces. Information interrelationships for complex situations suffer from the same problem. The user does not know what is missing or even how it should be assembled, and in fact, it can often be assembled in different ways depending on the particular circumstances of the user.

The design must clearly reveal the relationships because people have a hard time either finding or remembering to look for them (Kammersgaard, 1988). Mentally constructing the relationships impose a high cognitive load on the user. Unfortunately, creating the relationships in a manner easily perceived by readers can be difficult. Nanard and Nanard (1995) discuss a user's difficulties in perceiving the overall structure of a hypertext document, which essentially describes any web-based information system. Keeping track of location within a hypertext imposes a major cognitive load when the system structure and mental model structure do not match. As will be discussed in chapter 5, high cognitive load significantly impairs the ability to develop or maintain a grasp of the situation. Cognitive effort is minimized if the system design maps directly into a user's view or mental model of a specific task (Benbasat, Dexter, & Todd, 1986). *But more importantly than the information possessing a coherent structure, the users must be able to distinguish that structure* (Thuring, Hannemann, & Haake.1995) *and must be able to grasp the relationships.*

The fundamental problem many existing approaches is the lack of handling with information relationships. The systems contain the information, but lack a clear connections between information. The interrelationships that require a high salience are often not clearly evident and, complicating the design, can change as the user progressively understands the situation. Information within a complex situation has highly nonlinear relationships, some of which are not intuitively apparent. As a result, the nonlinear aspects of dynamic information means the analysis must consider rate of change and sensitivity to change (both direct and indirect) of various information pieces since "integrated design can only occur when all the real issues are allowed to exert an influence" (Casaday, 1991, p. 47). Failure to consider change can easily result in predictions that are consistent with the information, but hopelessly inconsistent with reality. In Figure 4.3, a person could notice the huge clouds created as part of the launch and decide there is a connection between the shuttle and the weather. Obviously, making such a misinterpretation would result in later information being misanalyzed.

FIG. 4.5. Space shuttle puzzle picture with the elements out of order.

As a specific example, consider how dominoes can be assemble, "as in a game of dominoes, there are logical relationships that require some pieces—some activities—to be arranged in a particular order, yet there are many possible orderings" (Constantine and Lockwood, 1999, p. 34). After a particular choice (game play) there are multiple possible options, but many others that violate the rules. Understanding the relationships (and history) that drive this fitting together allow for the construction of systems that provide integrated information and information relationships to the reader rather than a collection of information. To continue this domino analogy, the presentation and manipulation should start with a partially assembled game setup and allow the user to finish assembling the dominos to complete the game rather than starting with a pile of dominos dumped onto a table (or the shuffled shuttle images of Figure 4.4).

The design goal behind showing the information relationships "is to help people solve problems, rather than directly to solve problems posed to them (e.g., question-answer systems)" (Belkin, 1980, p. 134). The users must be made aware the information exists and have easy access to it, but, as we saw in Chapter 3, the overall system structure must allow them to choose what information to view. The informational system should provide access, not force the users to consider it. That's why the initial presentation must be a partial domino game and not a completed game, and most certainly, not a random pile of dominos. Presenting too much information would lead to cognitive overload, and trying to guide a person along one route through the process (i.e., using wizards) rapidly frustrates them because it forces them to view information they either already know or consider irrelevant, while failing to present the information relationships fitting the current situation.

### *Information coherence*

Making sense of the information and its relationships requires the information to be presented as a coherent whole. Campbell (1995) explain coherence in terms of connections within discourse as a way of ensuring the appropriate elements were clearly seen as related and connected. Text coherence exists at both the local and global level. Local coherence relations consist of the interconnections joining adjacent clauses and sentences. It is normally achieved through methods such as the use of words such as, because, but, therefore, and so forth. Global coherence is the overall textual coherence ensuring paragraphs and sections work together to create a unified whole within the text. Titles, headers, and topic sentences are markers of global cohesion. Texts must be both locally and globally coherent because text that are locally cohesive but lack global cohesion tend to inhibit comprehension and recall. Similarly, texts that are globally cohesive but lack local cohesion are sometimes difficult to read and comprehend (McNamara & Kintsch, 1996).

The large-scale coherence of the informational system comes from constructing the relationships between all the major pieces and lets the users synthesize the work as a whole (Thuring, Hannemann, & Haake, 1995). In a properly designed information system, users can rapidly access highly integrated information, work in an environment that allows for dynamic modification of content and format while maintaining the coherence, and receive support for examining related information, or cause-and-effect relationships.

Understanding comes when the user interprets the information with the proper mental model, which means the system must ensure the presentation matches the needs of that model as closely as possible. Murray (1995) qualifies the ideas of coherence by explaining that:

> It [coherence] refers to the way people try to make sense of what they read and hear, regardless of the actual explicit cohesiveness of the text itself. The degree to which readers or listeners successfully graph the intended meaning and relationships in the text depends largely on whether their expectations are consistent with the text (p. 18).

Lack of coherent information can be caused by failure to consider the user's understanding of the situation and the goals and information needs that are relevant. For example, the difficulties users have with Internet search engines or many intranets stem from a lack of coherence. Rather than receiving integrated, coherent information, the user is presented with a collection of documents that may or may not be relevant. Poorly design single source material can contain information chunks that lack any transitional ties, leaving the reader to make those connections. Although there is evidence a knowledgeable reader can imply the necessary coherence, it requires more cognitive resources and time.

## Information needs in complex situations

Much of the information structuring research attempts to predefine information needs and, thus, the system breaks down when users try to go beyond the solution envisioned by the designer. In a well-defined domain with closed-ended questions, the user approaches are limited and can be fully defined by the designer. In an ill-structured domain with open-ended questions, each user takes a slightly different path and the designer can't assume that an understanding of how one person performs the task describes anyone else. The designer can't even assume that the information needs are consistent between users or the amount of information a user will view before making a decision based on it. Real-world decisions get made based on choosing the first relevant solution and "muddling through" (Klein, 1999; Hollnagel, 1993). Yet, the designer is tasked with creating a design which provides the information when and how the user wants it.

The problem of overwhelming amounts of information is recognized throughout our society. Tebeaux (1996) quotes Al Gore as stating, "We are now drowning in information. We have automated the process of collecting information, but we have not successfully mastered the task of organizing and distilling the information for our productive use" (p. 40). Information overload discusses places two major obstacles in the path of reader comprehension: "the inability to locate what is relevant due to sheer volume, and overlooking what is most critical among relevant data" (Herbig & Kramer, 1992, p. 442). The definition of the real-world questions must reflect the users' information needs and psychological processes (Lansdale &

Ormerod, 1994). The method of dumping information upon users and hoping they can sort it out (the basic idea behind current World Wide Web search engines) fails because it does not contribute to efficient information processing. Web search engine studies show that most people don't use sophisticated search strategies, resist learning the complex systems and rules, and expect search engines to create effective searches automatically. Norman (1993) provides a nice summary of why people have a hard time mentally handling information:

> The power of the unaided mind is highly overrated. Without external aids, memory, thought, and reasoning are all constrained. But human intelligence is highly flexible and adaptive, superb at inventing procedures and objects that overcome its own limits. The real powers come from devising external aids that enhance cognitive abilities. How have we increased memory, thought, and reasoning? By the invention of external aids: It is things that make us smart. (p. 43)

Norman's external aids currently are the informational systems with their Internet connections tied to large back-end databases. So, although humans are highly flexible, the information must still be presented to allow for its easy integration and interpretation.

Simply put, people bring to the system real-world goals of obtaining information from a database to help solve a problem (Belkin, 1980)¯a goal which many systems fail to meet. Representing information with a graph, chart, picture, or video never truly solves the problem; only when it properly captures the rhetorical aspects of the information being conveyed will it have a chance of addressing the user's problem. The visualization mechanism must fit the information and convey the knowledge appropriately (Laplante & Flaxman, 1995; Tufte, 1983). Yet, as Foy (1996) points out:

> Business intelligence, corporate news, internal publications, organizational memory, external information sources, information content and quality, media literacy, information policy and security—these are the necessary ingredients of corporate knowledge management. More than ever, people need high-quality, timely information that is easy to obtain. (p. 23)

In an attempt to provide employees with the quality information Foy refers to, corporations are placing information on the Internet and intranets and building knowledge management systems. Placing information online to provide easier access and lower maintenance costs has been advocated for many years. Various research examined pre-web online report development within corporate environments (Doheny-Farina, 1988; Paradis, Dobrin, & Miller, 1985). There has also been research into effective presentation of online documentation (Horton, 1990). However, the online help research generally focuses on the learning-to-do format that supports data-entry. As information moved onto the web, both Nielsen and Spool, along with many others, have lead the discussion on how to present the information. Unfortunately, web research is often sidetracked by navigation issues, the relationship of text versus graphics, or equating ease of finding data with having meaningful information. E-commerce research is even worse with a focus on non-informational issues such as how to hold a person at a site longer.

### *Clear versus usable versus complete information*

Any discussion of meeting a user's information needs quickly hits on the issues of providing clear information, usable information, and complete information. Dye (1988) considers the clarity versus completeness argument to arise because a document that accurately describes the system often fails to be the most usable. He attributes the miscommunication to the different views of the designer/writer and reader. Lacking a clear understanding of what a reader wants, the writer provides everything, while a reader only wants information relevant to the current question. For example, older software documentation was that everything was written at the same level, regardless of whether the software command was a common one or a highly specialized one. As a result, the user had a hard time interpreting what was required (Martin, 1986). Unfortunately, achieving clarity, usability, and completeness is basically impossible in complex situations; the practice design issues become one of achieving a sufficiently high quality result across all three.

Simply providing access to the information (the old software documentation method) does not help, especially for addressing open-ended questions (Horton, 1993). User's faced with complex situations are asking questions that have no simple answer; the designer must work to give the user the ability to transform the massive pile of data into a useful package of knowledge. Even with high information content (a system often contains all the required information), because of poor design, the information cannot be effectively communicated. Think of the constant lament of helpdesk staff and technical communicators "the answer's here in the manual, here in the online help, here in the web site. What's the user's problem? Why do they refuse to read?" The problem is that with the information both hard to find and hard to process, the communication between the interface and the user has broken down and for all practical purposes the information doesn't exist (Bowie, 1996).

With clear information, the user can easily grasp the relevance of how the information applies to the current goal. Usable information, besides being relevant, can be easily tweaked or manipulated to fit the overall goals. Complete information is harder to define. In a very simplistic definition, it would be all the information. But that can easily overload the user. Limiting the definition of complete slightly to be a complete set of relevant information for the current goal gives a workable definition. The difference here is can be seen by thinking about some software manuals that provide full descriptions of every available field and button when explaining a task, when in normal use only a couple are used. The full descriptions give complete information, but the clarity and usability suffer because the user must figure out what to ignore. And then, of course, there is documentation that is very complete but written "by and for techies" with complete disregard for other audiences.

Clear and usable information also includes the idea that readers want fast access to information. Fast access often means deciding what is directly relevant to the person's goals and information needs and what is "nice to know" information. A reader who, after skipping text, becomes confused and frustrated the first time will not return a second time. The information must be presented in a consistent form appropriate for the information needs. Of course, across closely related complex situations, what qualifies as required and as "nice to know" varies depending on situation specific factors. Many times, with poor analysis, a design presents too much information and lets the user figure it out (Figure 4.6). With a clear vision of user goals and a dynamic system to support implementing that vision, only the relevant information can be presented.

While it may seem obvious that information needs to be in a usable format, in final designs, it often becomes confused with providing complete information. Usable information cannot not be prematurely defined as an information containing a complete set of information. That's a problem with many canned reports; lacking a clear understanding of how the report gets used, it contains all information remotely related to the subject sorted in the overarching hierarchy. Fortunately for the user, the modern online analytical processing (OLAP) systems are providing more effective ways of dynamically viewing and sorting information.

Issues of completeness rapidly meet up with the real-world issues of the impossibility of providing complete information. Real-world complex situations essentially always consist of working with incomplete information. Time and expense prohibit the system from capturing a complete set of information. People come to the informational system lacking a complete mental picture and expect the informational system to provide that picture. However, they generally accept as a given that the information is incomplete and do not search for an optimal answer, but rather are content with finding a sufficient answer based on accessible information (Klein, 1999; Orasanu & Connolly, 1993). Because people cope with incomplete

FIG. 4.6. Present only the useful information. If a person only wants information about the shuttle itself, the rest of the information is noise and distraction.

information in different ways, they need the capability to solve the problem in their own individual ways, especially with the "fuzzy" success criteria of many real-world situations (Hallgren, 1997).

Meeting the design goals of clear, usable, and complete require information at the right time and in the right order. It also allows the easy expansion and contraction of that information to match the person's immediate information needs. At no point does this necessarily translate to providing all the information immediately available. The important element is accessible information, not sticking it right in front of the reader. Information dumps lead to cognitive overload which makes the information less usable. The amount of information a person needs out of the total available versus the usability of that information follows an inverted U-graph (Figure 4.7). The perfect system presents the information which places the reader at the peak of the curve. Unfortunately, even identifying all the factors which contribute to defining the peak, much less defining the presentation mix, is almost impossible. However, although designing an optimal system is basically impossible, by remaining conscious of the curve when designing the information, the reader can stay close to the top part of the curve.

In clearly presented information, the reader understands the relative importance of the contents: The salient information relevant to the current situation and current goals clearly stands out.

Unfortunately, the literature on presenting information consistently lacks an empirical focus; it consists mostly of anecdotal evidence based on experience, and appears in the form of interface guidelines which are not positioned for extensibility (Hackos, Hammer, & Elser, 1997; King & Teo, 1996; Rainer & Watson, 1996). These guidelines provide a good start, but lack the foundation to understand what to change and the effects of any change when the situation is different. Fitting the problem to a person's mental model needs more than the consistent "look and feel" of conventional interface guidelines, but rather it must minimize mental transformations (Hix & Hartson, 1993).

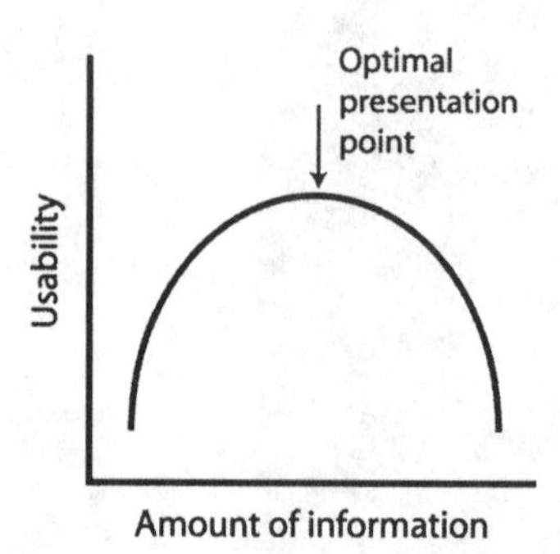


FIG. 4.7. Amount of information versus usable information. The optimal information presentation happens at the top point of the curve. Presenting too little or too much lowers the overall usability.

Whereas violating accepted design guidelines often renders a design unusable, strict adherence carries no assurance of creating a useful design for addressing complex situations. In general, too much information design fails to address or connect the design to why the user needs the information and to the rhetorical aspects that maximize the communication. Consistency is a double-edged sword. The problem with an extreme focus on consistency is that it often produces consistently poor interfaces. The key is to determine what are the best, most appropriate, and most usable techniques and designs, and implement them consistently.

## Privileging complete information

I've talked to some technical writers who's main concern was with making sure their documentation complete and correct. Then, if they had time (insert laughter), they would worry about making it look nice and general usability issues. An interesting perspective on writing: Making sure every nuance was documented was the most important thing and everything else was a distance second.

I've also noticed that these writers tended to be writing programmer level documentation for APIs or system implementation. They are the people tasked with writing those poorly formatted tomes of minutia on every possible element of the programming toolkit. The result of course, is that the programmer gets used to such system and sees no reason why everyone can't learn to use it too. As a result, those tomes become the model for later systems which are designed and perhaps even written by the programmers. If as an audience group programmers want documentation written and formatted that way, fine. However, when it starts to carry over to the interface and how information gets presented to the general user, then there is a problem.

### *Situation relevance*

The previous section described how completeness (defined as *all* the information) is not a necessarily good thing. Rather, clear and usable information depends on limiting completeness to information that is the relevance to the situation. The design goal should be providing a complete set of information that is relevant and removing all information that is not relevant, and then arranging it in a usable manner. The fundamental principle guiding the information designer should be providing access to the information in a form that enhances the reader's awareness of the situation and which fit the (business) rules of the game.

Consider that many people access the information, but all require it at highly differing levels of completeness and organization. The users' needs and goals must drive the problem, the standard data drill-down found in many systems assumes everyone wants the same hierarchy of information and organization with different detail levels. This assumption is based more on simplifying the implementation than on efficient transfer of knowledge. Standard drill-down techniques rely on a fixed hierarchy. The pros and cons of using a hierarchical display of knowledge is not the issue here; rather it's the idea that each major user group (if not each person) requires a slightly different hierarchy (Brehmer, 1988). The designer's challenge is to figure out how to provide that slightly different hierarchy in a manner which still allows the system to be developed and function. Only recently with tools such as XML as it become possible to even attempt to create dynamic hierarchies customized for each user.

The analysis challenges involve issues such as:

- What is the minimal information required to understand this situation? Although a user does not need complete information, incomplete information with respect to the situation is highly detrimental. For example, I just want to know that the three lights on my printer mean, I don't want to have to dig through the entire printer manual to find out.
- What other information might be required to support the minimal information set? This could be things such as examples, definitions, longer explanations (what could be clear for one group is too concise for another), or more comprehensive explanations (one group needs a higher level or detail or more technical explanation).
- How does the information interrelate? Information gains it situation relevance based on how it relates to surrounding information. Clearly laying out these relationships helps the person grasp them. Only with dynamic systems can these change as the user needs change.
- What order does the information get used? To be relevant, it must be presented at the proper time. Out of sequence information can be more detrimental than helpful.

### *Information salience*

Information salience follows directly from information relevance. As a basic definition, salience refers to giving each information elements the prominence and importance it deserves. The most important (highly salient) information should be clear and easy to find, the less important information should never distract from more important information. The sidebar in this section shows how many software user manuals suffer because they don't consider information salience, but treat all information at the same level. Without presenting information according to its salience, large quantities of unimportant information can dominate a display and make the important information hard to find.

Consider what happens if the available information gets focused on a single aspect; the total picture becomes distorted. Figure 4.8 shows a space shuttle launch but with the various sections highly distorted. The user gets a distorted view that does not really represent how the shuttle looks. It would be easy to prevent this type of distortion if we could actually present clean, clear images of situations, but normally such visuals don't exist. Then, the available information can take over and dominate or bury the important information.

The distortion on the shuttle is obvious, but for most situations, the distortion is much less clear. How can a person who doesn't understand the situation know when the information is distorted and which data is distorted? When the person does not know, Wickens and Carswell (1995) developed the idea of the proximity compatibility principle (PCP), which is similar to gestalt proximity concepts. PCP states that the perceptual characteristics of a display should be designed so that the information arrangement fits the cognitive demands of the task. In other words, information which is used together should appear close together. If the task requires two set of information to be integrated, the system perform the integration and display the results. Conversely, information that is not used together should not appear together. Notice how this principle pulls together information relationships, relevance, and salience. On the other hand, although this idea may appear self-evident, many systems violate it.

A system that has high information salience means ensuring the user looks at both the proper information and the proper amount of information. However, this provides the designer a major challenge because the user rarely know what to ask for. A gap between the user and system model causes this deficiency in many current systems (O'Malley, 1986). The design of a system that truly places information within the situational context must be "so proximal that the user's flow of creative thinking is not interrupted" (Basden, Brown, & Tetlow, 1996, p. 157).

An important measure of the difference between a well-designed and a poorly-designed system for complex situations is user's ability to gain an understanding of the situation. Unfortunately, many usability tests focus on too low a level of situation context. Counting clicks or timing how long it takes to find a single piece of

FIG. 4.8. Space shuttle launch with various parts of the image distorted. Most information a person gets is often closer to a distorted view than a sharp crisp picture. But, unlike a space shuttle image, they don't what is distorted.

information are both examples of tests concerned with very low level issues of understanding. This type of test only focuses on if the user can find the information, rather than if the information relationships are clear enough that they can be synthesized into a clear overall understanding. Stopping with this type of testing runs the risk of optimizing a system for presenting single data points and not presenting integrated information. Understanding a complex situation requires integrating multiple pieces of information and acting on them, not rapidly finding single pieces. Part of the measure of a design being proximal comes from the ease of accessibility of information with respect to meeting the user's goals. Although counting clicks can be a good early test, the real usability comes from testing that the interface becomes an important part of the understanding process as the user assimilates the information through all the means provided in the interface (Treu, 1990).

Information salience is important because having access to different information can change how users frame the problem and how they interpret the information. Communication is inherently difficult because we must imagine how the ideas and concepts relevant to our problem might be represented and were they might be placed within the system. This can be doubly difficult because we lack a clear understanding of the problem (Marchionini, 1995). "There is considerable evidence that a person's manner of characterizing a situation will determine the decision process chosen to solve a problem" (Endsley, 1995, p. 39). Roberts (1989) found that, with impaired understanding, people sought more clearly defined relationships within the existing information or pointed out ambiguity, whereas if they simply felt their understanding was unsatisfactory, they looked for more information. A problem with many existing (and failed) decision support systems was that the designers applied a technological solution that provided a low level of situational understanding, but never defined the requirements to ensure the user understands the relationships. Post hoc error analysis reveals that most interactive system errors can be traced to the fact that the data does not accurately represent "real world" conditions (Macaulay et al., 1990). The system provided the data, but didn't provide adequate context to allow the user to create a comprehensive picture.

Waern (1989) points out the design problem is one of perspective. Development people only want to work with data which can be placed in a database and retrieved as a ready-made answer. Users want information which does not fit into such nice clean buckets. An effective information system must support the changing information needs and not focus on only one (Elam & Mead, 1990; Shneiderman, 1992), especially as users constantly adjust the plans to reflect changing conditions and information (Jirotka & Goguen, 1994). Woods and Roth (1988) define the critical question as "how knowledge is activated and utilized" (p. 420) The question concerns not merely whether the users know some particular piece of domain knowledge, but do they understand the relationship between different pieces of information. Do they know "that it is relevant to the problem at hand and does he or

## User guides and lack of salient information

A common complaint about software documentation, especially vendor provided documents, is that information is hard to understand. A major part of the problem is that the important information does not get the presentation it deserves. Instead, all the features of the software are documented using the same template and to the same exhaustive level of detail. The 10% of the features that get 80% of the use receive no more emphasis than the features only used infrequently by the system administrator.

Part of the problem can stem from legal issues of ensuring the entire product is documented, but the design problem seems deeper than that. Do the writers actually understand how the software will be used? And do they try to write to support accomplishing the most frequent tasks? Considering the number of software manuals that consist of starting with File-New and completely defining each field in the dialog box, I'm not sure I really want to know the answers.

It seems like most people prefer the third party books on the major products. These books often include much less detail than the vendor manual (think of the success of the *Dummies* books), but they place the important information in context. Rather than working to provide all the information, they focus on situation relevance and ensuring the important information is salient.

she know how to utilize this knowledge in problem solving" (Woods & Roth, 1988, p. 420). Users require information that relates to the overall situation and they need to understand that relationship (Robertson, Card, & Mackinlay, 1993). The information design must work to ensure that relationship is captured in the early analysis and made as explicit as possible in the resulting design.

### *Signal to noise*

The signal to noise ratio is a common concept in radio communications and electronic communication in general. For a radio, the static is the noise. Too much static and the storm report gets drowned out, or at least you must listen closely to understand the announcer. Electrical engineers working on a voice circuit can easily define what are the signal and noise elements. The original voice spoken into the transmitter is a known quantity and anything else that comes out the speaker is noise that should be eliminated. Although actually eliminating the noise may be very difficult, electrical engineers have well-understood techniques such as bandpass filters to employ to reduce the noise.

In an information system, "signal" consists of anything that the user needs to know but doesn't, and "noise" is anything that they don't need because it is irrelevant or already known, and thus, is just cluttering up the screen to the detriment of the signal (Figure 4.6). Tufte (1983) was referring to essentially the same concept when he discussed data/ink ratio and chart junk. Basically, noise consists of all the stuff that interferes with communicating what the person needs. (One anonymous reviewer questioned why I didn't discuss signal theory here, but I feel that it's too abstract to contribute to the discussion of this book. In other words, as far as this book is concerned, it is more noise than signal. For those interested, Wickens (1992) has an excellent presentation.)

Unfortunately, information designers do not posses such a clear cut set of techniques, such as a radio circuit band-pass filter, available to electrical engineers. For information systems, taking the raw data in a system and deciding what is signal and what is noise proves to be extremely difficult. Because of the different goals and information needs each user brings to the system, there will never be a clean method of resolving the signal to noise ratio. Rather, the extensive analysis and testing all help refine the understanding of which is signal, which is noise, and most importantly, which contextual situations cause noise to become signal and vice versa.

The difficulty of resolving the signal to noise ratio occurs at many different levels. For example, it is reflected in the basic ideas of the differences in writing for novice and experts or writing for high and low literacy levels. The extra explanation required for the novice becomes noise for the expert. And of course, the reverse can occur: without more detail, the written-for-an-expert text degenerates into noise since it can't be understood by a non-expert. It also occurs in the fundamental information requirements for a job. The technician needs specific details about the task while the supervisor wants a higher level less detailed view. As a result, signal for the technician becomes noise for the supervisor. Interestingly, readers with a high knowledge level in the general subject tend to have poor comprehension of text written for a low knowledge reader (McNamara & Kintsch, 1996). The extra detail bogs them down and results in less focused reading, resulting in missed information.

The signal to noise issue enters design for complex situations because people tend to ignore what they don't understand. Rather than working to understand the entire block of text, they reduce the cognitive effort and ignore the parts they don't understand. Such is human nature, no design can change it; instead, designers must acknowledge it and develop designs that try to minimize the effect. People are given too much information all day long and the majority gets ignored, thus it's very easy to ignore any information that isn't understood. On the other hand, people can pull information from a very noisy source if they know what they want and are familiar with the information they seek. For example, listening to a song on car radio when the window is down. Or laboriously extracting bits and pieces of data from multiple reports to gain understanding of a production problem. Mentally separating signal from noise requires an expert's viewpoint, someone who differentiate the signal and noise. People with lower knowledge levels can't perform the task because although they can see wants in the reports, they

### Dealer information screen

One customer relations system I worked with dumped everything about the dealer onto the screen. Considering that the main caller to the customer service people were the dealers, filling the top half of the screen with address and phone number information was accomplishing nothing but providing noise. Dealers know their address and phone number. Instead, put the information a customer service person normally needs: account status, and so forth. on the top half of the screen.

can't make the necessary connections to form meaningful relationships. And most people using a system are not experts so they require help from the system in developing a clear view of a situation.

Maximizing the signal means understanding both the user's terminology (versus system terminology) and the user's situation. Writing in user terms and avoiding jargon has been a standard guideline for many years. Quite simply, anything written in unfamiliar terms will be considered irrelevant and ignored, regardless of its importance. But more than just the proper terminology, the text must fit within the situation. The focus must be on presenting information in the formatted both expected and understood by the reader. For example, medical information and business balance sheet have formats fixed by long tradition. Changing the format could actually result in someone ignoring the entire page with a logic of "this doesn't look like the XX report, so it doesn't have the information I want.

### Signal to noise in screen clutter

The design technique of piling information on a computer screen assumes a person can pull the signal out of all the noise. Consider the uncluttered appearance of Google is versus Alta Vista, Yahoo, and most other web portals. Alta Vista tried to move from being a search engine to a one-stop everything and often ended up looking like a beach front t-shirt shop. A bright flashy collection of wonderful looking knickknacks that descend into horrid sameness after about three visits (to the store or the portal). Finding the text entry box can be a chore.

The cluttered appearance can arise because without an in-depth understanding of the problem and the user's goals and information needs, the designs and developers often are forced to take this approach because they don't know what the user needs. Lacking a clear understanding of what information is relevant and how it interrelates, all the information gets treated as equal.

## Signal to noise with Microsoft Word styles

Consider the amount of information that different people need to act on the direction "In this document, change all Heading 1 styles to 20pt Arial Black with 18 points of space after."

Present that directive to three different users.

**A person who understands paragraph formatting and styles in Word**. The statement is self explanatory and takes about 30 seconds to accomplish.

**A person who understands styles in Word, but not paragraph formatting.** The difference between this person and the first person is that he doesn't understand the part about the 18 points after. As a result, the change to 20pt Arial Black is understandable. But two different actions could occur with the 18 points after. The person could either accept it as part of the direction and try to figure it out or it gets reduced to noise and is ignored.

**A person who does not understand Word styles.** This person could not act on the direction at all. As a stand alone direction, the person might work to figure out what it means. If the direction is buried in many other directions, there is a good chance it will get ignored. The word "style" and the general concept are not part of the person's knowledge, so it is ignored. Most likely, the person would just manually change the heading format for each heading in the document.

If a system was designed to support these three people, it should allow for dynamic presentation. Whereas the first person would not use the system for this question, the other two need different information. One needs information on how paragraph formatting and the other needs information on styles. Considering this as a complex problem, both people need an integrated presentation that discusses formatting and styles in the proper relationship and amount. The standard documentation method of discussing both paragraph formatting and styles as a stand alone topic does not contribute to the user understanding how they interrelate and how to accomplish this task.

### *Importance of disconfirming information*

If most people are asked what type of information they consider most salient, they answer something along the lines of "information to show I'm making the right choice." Interestingly, the information people need is *not* the information which shows they made the right choice, but rather is information which might reveal they made the wrong choice. The importance of disconfirming information for information design comes from the way people make choices. In most cases, all of the factors will not be viewed. Rather, the user has a solution in mind and will only look for confirming information. As the amount of information increases, users increasingly ignore information incompatible with the decision they want to make (Ganzach & Schul, 1995). Because people look for confirming information, not disconfirming information, specific elements that can provide disconfirming information must be uncovered in the analysis and given a high salience level. Although human nature makes people want to ignore them, the design must make it as difficult as possible to ignore them.

## Disconfirming information

Table 4.1 shows various scenarios for when a problem-solver has options 1 and 2 and various factors that help distinguish between them. An important item to gain from looking at the table is the way supporting information can be provided (factors A and B are both true), yet the option choice could be wrong.

As A and B are correct, option 1 must be right! Only later does the audit reveal what wasn't looked at. Thus, the information designer must ensure that the disconfirming factors of E, F, and G are prominent in the information presentation.

Table 4.1. The decision driving information must be what disproves the option.

| Decision choice | True if choice is true | False if choice is false | Not relevant |
|---|---|---|---|
| Option 1 | A, B, C | B | D, F, G |
| Option 2 | A, B, D | F, G | E |

| Assumption | Factors that are true | Have enough factors been considered? |
|---|---|---|
| Want option 1 | A, B | Unable to decide as C and E are not considered. There is insufficient information. |
| Want option 2 | A, B, C, D, F, G | Unable to decide as E is not considered. But this seems to have an abundance of information available. In many cases, a person would see that A, B, and C exist and go with option 1, and not realize (or remember ask) that since E = false this is the wrong choice. |

## *Presentation*

This section touches on some issues to consider when presenting information to a user. Because chapter 6 discusses system operation and chapter 7 discusses putting everything together, this section may appear somewhat sketchy. The purpose is not to explain how to design a presentation system, but, rather to present concerns that can interfere with the user obtaining high quality information.

The presentation influences how the users perceive the information relationships (Johnson, Payne, & Bettman, 1988) and becomes the crucial factor in maintaining the flow of information to enable users to achieve their goals and maintain situational context (Laplante & Flaxman, 1995). The design must consider the presentation and use of information as a method of fitting it to the information needs contained in the user's mental model. (Robertson, Card, & Mackinlay, 1993). The design can help reduce cognitive overload because, with effective cue words in the link text (definitely not nebulous labels such as "Goal"), the link text itself serves as a reminder to users about the information contained behind it. For example, a label on the screen could be "Student concerns" or "Maintain student focus" with both linked to the same text. However, the latter serves to remind the user of a of the specific information available behind the link, especially if it was previously viewed. The former could be placed on almost any screen and does nothing to mentally stimulate information retrieval of both the individual item and its information relationships (Einstein et al., 1990).

Issues of concern for effectively presenting information:

- People evaluate information based on the order they receive it. Information presentation must be in the order that best helps gaining a clear understanding. Systems that strive to always provide complete information violate this because people then have trouble looking at the information in the proper order. This also explains some of the bias that occur of people liking or comparing everything to the first item they saw, regardless of whether it actually fits their goals.
- Salient information must have proper presentation (Basden, Brown, & Tetlow, 1996). Potential problems can occur when minor, but easy to present, information occupies an excessive amount of the display area and, subsequently, diminishes the salience of the important, but harder to present, information.
- The wording influences user response. Word choice forms an important, but often ignored, part of user understanding; word choice can greatly influence how certain users feels their understanding of the situation.
- Users tend to change their assessment strategy to fit to the presentation method, rather than transform the information to fit a better assessment strategy (Johnson, Payne, & Bettman, 1988). This is another way of saying that people change their work habits to fit the system. Rather than properly evaluating information, people tend to take what the system gives them.
- People think they use more information than they actually do. They tend to underestimate the weight placed on important cues and overestimate

weight placed on unimportant cues. Also, experts are just as susceptible to this as novices (Andriole & Adelman, 1995, p. 25) .

- People suffice; they only look at information until they feel they have an understanding of the situation (Klein, 1999; Simon, 1979). Unfortunately, they may or may not actually have a seen enough information to give them a complete picture.
- Viewing complex information is not a straightforward, linear process. People gain an understanding of the situation by following "a long and recursive process with backtracking and erratic switching among the following activities: thinking about ideas, production, reorganization, modification, and evaluation" (Nanard & Nanard, 1995, p. 50).
- Information relationships must be clearly presented. People are very good at seeing interconnections and grasping the big picture, they are very poor at correlating and analyzing large amounts of data. Computers have the opposite problem. Systems for complex problems must be designed from the ground up to allow both people and computers to do what they do best.

The problem with ineffective information presentation is that, from the psychological standpoint, the users suffer from poor information dissemination. Mentally processing a collection of data is highly inefficient with high cognitive workload. In poor design, extracting the relevant pieces of data and converting them into information gets pushed onto the user, rather than being performed by the system.

## Film processing pricing example

As part of a customer service application for a wholesale photofinishing company, if the customer service person entered the order number, they could see the order pricing information (sort of). Instead of giving the prices used to price this order, the system dumped all the information relevant to that order type.

For example, consider a 24 exposure 35mm roll with single prints. Pricing was complicated: base price, region price, multiple coupons (percent off, fixed amount off), national chain specials, and multiple local specials. Thus on the customer service side, you would expect the actual pricing information used to highlighted. Instead, it showed all possible prices and never flagged that a 10% off coupon was used.

Major informations problem. All the prices are not initially relevant. The customer service person needs to know what pricing method was used. The other prices are extraneous until they determine the wrong pricing method was used. Then the other prices need to be easily displayed.

This display provided information not relevant to the situation and, by not marking the used pricing method, failed to highlight the salient information. With all the prices displayed, the signal to noise ratio was too high for the customer service person to easily sort them out. (Several major design errors such as this one resulted in the system being cancelled after 18 months of development but before deployment.)

## Long example—E-commerce

This example considers how a consideration of information needs an be applied to the design of a e-commerce web site.

In both of these examples, the site design must support two separate sets of goals. The business goals of selling products and the user goals of being able to find and buy a product. Whereas it some people might consider these the same fundamental goal, they can be strikingly different. As one example, think of how a grocery store very carefully and intentionally puts high volume items at the back which hurts helping the customer buy a product, but increases sales, because by making customers walk through the isles, it encourages impulse buys.

In both examples, we see that the design does not support the complex situation the user brings to the site, but continues the traditional printed catalog method of presenting information.

### *Cell phone accessories*

As an example of non-integrated information, one major cell phone provider site using a totally disconnected method of operation. About after you log in to manage your account, you see the "Welcome, [insert name]. You own [make and model number] phone. Buy accessories?" Clicking on the "Buy accessories" link takes you back to the non-logged in portion site where you need to start hunting down accessories for your phone. The phone make and model does not carry through and any previous accessories you purchased through the site does not carry over. Instead of making the person start from scratch, obviously the phone information should carry over so only applicable accessories are displayed. But it should go further, and suggest new accessories to compliment existing ones. It's a waste of display space to show a headset when the person already owns one. Also, a site where the user's model is known should never display items with the a line like "fits models 92, 93, 95," especially when the user has model 97. That's ok for a printed catalog, but not for a site that presents integrated information. (Of course, sites that say that, don't present integrated information.)

### *Women's apparel*

As a second (more extended) example, consider the design of most women's clothing web sites. Current designs mimic the large discount stores which leave shoppers on their own. Each type of clothing (blouses, skirts, dresses) has its own link off the home page. After deciding on a blue dress, the user decides to buy new shoes. The shoe page displays all the available shoes with no connection to the dress which was just bought. Likewise if the user decides to get new jewelry to wear with the dress.

In contrast, think of the person going into a major department store. One salesperson would help pick out the dress, walk over to the shoe department and only suggest appropriate shoes, and then do the same in jewelry. If an e-commerce site mimicked the department store, it would provide help in selecting matching items and suggest other related purchases. The discount store considers purchases as a simple situation whereas the department store understands shopping as a complex situation.

The question becomes of why the web site does not do the same thing. Consider the possible interaction. Once the person makes an initial purchase of the dress, if they indicate they want coordinating shoes, then the first shoes should match both the dress, the season, and the user profile (the shoe tastes of a teenager may differ markedly from a 30-something). However, all the shoes must be easily available in case the person doesn't like the first set.

The first site design assumes a simple situation with a linear flow that has each purchase independent of any other. The user decides on one product, places it in the shopping cart, and repeats for each item. The second design acknowledges both that purchases can be related and that people with highly specific, if varying, tastes are involved. The people create the complex situation as they look as the dress and available accessories in a multidimensional way. Exactly what they will pick out cannot be predetermined, but by working in general categories, the system has a good chance of presenting acceptable choices.

## *Designer and writer issues*

Chapter 7 describes a set of steps to consider when designing for a complex situation. This section looks at those steps as they apply to supporting e-commerce.

### Scope

Having a clear set of scope statements if important. The first example worked like a discount store and the second tried to imitate a major department store. Rather than a scope of "sell clothes" the scope statements must clearly define how the site intends to sell clothes.

### User groups

Knowing the customers is paramount to success for any retail store.

- User groups have different views. Teens are different from older women and will match different shoes and jewelry styles although they may buy the same dress as an older customer.
- The same user group wants different accessories depending on type of outfit or intended use. Was the dress mostly appropriate for business, formal wear, or weekend club hopping? All of these require different accessories.

### User goals

Besides the obvious user goals of buying a new dress, there are other more subtle goals. These include information goals that supports the purchase decision, such as type and weave of the material or place of manufacture.

One way to uncover the user goals for this site would combine ethnographic studies with a good set of personas and scenarios (marketing probably already has a good start on this information). The ethnographic study needs to capture how an experienced store clerk interacts with various customer groups. Interactions with the user groups would also allow basic metatagging that relates clothes and accessories to match current fashion trends of each group.

### Information needs

The user goals drive the information needs and define how deep and detailed the potential information needs are. The first site description ignored user information needs and just presented various products. The second site worked from the current purchases and presented appropriate products. Knowing what those appropriate products are requires understanding the user needs.

The design team walking the personas through sets of purchase scenarios would ensure the information matches the persona needs.

### Information relationships

The second site tries to match the information needs (what shoes will look good with this dress) and present that information. Consider how the different options and clothing accessories play into different needs? Maybe the shoes must fit the dress, but must also be wearable with a variety of outfits. On the other hand, shoes colored to match the dress could be a major consideration for a formal dress.

### Presentation

Balancing the information needs against the user groups is major difficulty. Some type of profiling is needed to ensure the information fits the user group. Also, the overall tone of the presentation might change: different color schemes or model posses for teens versus older women. A dynamic system can provide the different accessories and even change the presentation style.

# 5 People

*The great obstacle is not ignorance but the illusion of knowledge.*
*—Daniel J. Boorstin*

*People's behavior makes sense if you think about it in terms of their goals, needs, and motives.*
*—Thomas Mann*

Thus far, this book has looked at the user's goals and information needs without directly considering the cognitive factors that drive how a user views and responds to a situation. However, as the volume of information increases, considerations of people's cognitive processes become paramount for providing effective design for complex situations. Previous chapters considered how the information was defined and displayed, here we look at cognitive aspects of how a person interacts with a complex situation and interprets relevant information. This chapter looks at the user and considers the following:

- How designers view the user and what type of elements go into defining the levels of user knowledge.
- How cognitive resource limitations influence people's responses to information and gaining an understanding of a complex situation.
- How different users go about understanding situations
- How users use existing mental models and how they develop new ones.
- How people respond to information and what factors can influence these responses.

## Introduction

With human nature, rather than using highly logical processes, most decisions are made by choosing the first relevant solution and "muddling through," (Klein, 1999; Hollnagel, 1993, p. 32). Unlike computers, people are non-deterministic; they produce varying output in response to the same repeated input. Although there may be a single "correct" response at any one time the actual response will vary, deviating around the "correct" response to some degree, due to unpredictable biases and both systematic and random error. When complex situations are considered, with the wide variation in goals and information needs, and the lack of a single path, the variations between users can cause major headaches for the designers. Headaches that cannot be ignored because they stem from cognitive factors that influence how people interpret and understand information. In other words, the biggest headaches for designers can be the biggest drivers of how effectively a person can use the system to understand the situation.

A fundamental concept in defining the users' goals and developing an understanding of how the users' form the connections between a situation to applicable information (Kammersgaard, 1988). For a design to accomplish this requires placing major emphasis on acquiring and using knowledge of the domain, the users, and their tasks and generally supporting the users' psychological needs, rather than just meeting technical requirements (Bowie, 1996; Hefley, 1995).

Kieras and Meyer (1998) point out that "humans choose task strategies that incorporate optional features that are not based on either architectural constraints or task demands. These optional features strongly influence performance, but cannot be identified by conventional task analysis methods" (p. 2). This break down between the designers' and users' thought processes explains why conventional task analysis works for well-defined domains but fails for the ill-structured domain of complex situations. In a well-defined domain, the user approaches are limited and can be fully defined by the designer. In a complex situation, each user takes a slightly different path and the designer can't assume that an understanding of how one person's performance describes anyone else.

Thus, addressing the user aspects of a complex situation requires a thorough analysis of the task at hand, coupled with psychological studies of characteristics of the users and an understanding of the situational context. Together these offer the most enduring information for information designers. Provided these processes are described and understood "in terms which do not depend upon technology, their implications will endure compared with the ephemera of technology" (Lansdale & Ormerod, 1994, p. 18).

### Cognitive resources

Although many factors may influence the interpretation of the information, the systems of interest for this book contain at least one human doing the interpretation.

With at least one person involved in interacting with the situation, the design must account for the associated human cognitive processing (Benyon, 1998), a processing method which forms both the strongest and weakest part of the overall interaction. It forms the strongest part because people can innately form and grasp the relationships which give rise to understanding a complex situation, while most computer system have not progressed beyond understanding very limited situa-

### The problem of technical, but no people, requirements

I've worked on too many projects where the people objectives never were defined. The technical requirements carefully detailed the under-the-covers program operation and technical considerations to enhance system performance.

One such project was a billing and invoicing system that would be used in over 50 company locations across the United States. The system analysts dismissed defining people requirements by claiming each plant was different and we (internal MIS) couldn't tell plant management how to operate their plant. The operational problem was that the people doing the price maintenance could be from, depending on the plant, customer service, marketing, or production. Also because of time and financial considerations, we couldn't directly talk to the real users; we talked to regional coordinators. The coordinators understood what the system had to do and could relate it to the business goals, but they only understood operations from the point of view of the one or two plants they had worked in. In the end, the interface design did nothing to support different users; it had a single design which directly reflected the underlying database structure. The system as programmed worked, however, it made no allowances for fitting the user's view of the work. In a fitting ending, once plant management saw the system, they all claimed their people would never be able to use it. Because of the usability howl they raised, the project was cancelled. And one the lead programmers told me she couldn't understand why because "the system technically worked and worked well."

tions (a reason we don't have the artificial intelligence found in science fiction books). It is also the weakest part because the human mind has strong biases and severe limitations on the amount of information it can process without being overloaded or without dropping information. An understanding of this limitation comes from research which has defined human's cognitive resources.

People seem to have a fixed-size pool of cognitive resources which they allocate to all mental tasks in which they are currently engaged. Depending on the specific researcher and how the theory was constructed, the amount of cognitive resources and how they get allocated to tasks can vary, but across all models, the actual amount is very low. Unfortunately, this cognitive resource pool seems to be extremely small, which results in it being easy to exhaust. Miller's (1956) well-known work on 7 +/- 2 items was early research into the size of the cognitive resource pool. Each mental task consumes resources, especially when the tasks must be done in parallel. Once the resources are exhausted, the user suffers from cognitive overload which leads to a high number of errors or skipping information. Besides jokes about not being able to chew gum and walk at the same time, limitations on cognitive resources explains the difficulties people have simultaneously performing tasks such as driving and talking on a cell phone.

The cognitive load associated with any particular situation has two components: The intrinsic cognitive load and the extraneous cognitive load.

| | |
|---|---|
| Intrinsic cognitive load | This is the amount of cognitive resources consumed by the task. Depending on the difficulty of the task, this value may vary. In general, for any one person, this value is fixed for a task and can't be changed. For example, the amount of cognitive resources required to analyze a report is inherently higher than filling out a timesheet. Because the values are inherent in the task, this value will not change with different presentations. |
| Extraneous cognitive load | This is the amount of cognitive resources consumed by the extra tasks involved in performing the basic task. Examples include tasks such as manipulating the windows and files of a software application, remembering data location, and using a particular tool to manipulate data. This value can be changed and the focus of usability and information accessibility is to decrease this value. The difference between an easy to use and a hard to use application, when both accomplish an identical task, is often the extraneous cognitive load imposed on the user. Providing more efficient data manipulation tools or better information presentation both reduce the value. As the major element of the overall cognitive load associated with a task which can be changed, it must receive special focus during the design analysis. |

### *Interweaving cf tasks*

The design of too many systems assume a straight forward approach with the reader starting, working straight through, and reaching an answer. Many design methodologies also work from this assumption of working straight through the task. But an information system for complex situations is not a typical system for which design methodologies were created, as Chalmers (1999) points out.

> The traditional computer science approach as giving insufficient attention to the intricate details of how information on an individual's desktop is interwoven with the rest of that person's working environment of people, institutions and cultures. Such concerns are largely absent form the practice and theory of data retrieval. (p. 56)

Regardless of how people describe their task performance, research shows they tend to use an opportunistic organization and interweave tasks (Cypher, 1986; Visser & Morals, 1990). In studying how people use computer programs, Cypher found many people do not follow a linear progression; instead, they interleaves multiple jobs. Some are interruptions that must be carried out immediately (Jack runs into the office with a quick question). Others are of the "while I'm at it I'll do this other task" type of jobs where they stop one task and perform another simply because it is convenient to perform it right then. In either case, the person does not attend to the situation start to finish, but they want to be able to easily pause and later return to the point at which they paused. If they paused for very long, they may forget some of the situational factors. So, besides simply letting them pick up at the stopping point, they need a method of quickly reviewing the situation.

Capturing the task interweaving presents a substantial problem for the analyst. At the minimum it requires observations of real users done under real conditions. Simply sitting down and having the person show how they perform a task will only capture the basic task and probably in the "company approved" manner, which may have little resemblance to how the task is really performed. How to capture these requirements is a complicated issue: how do you realistically observe people using a complex information system? They would be asked to search for and integrate information which they currently don't really want. Think about it: You were asked to read information on hiking dangers in the forests of Peru; although you like the idea of that vacation, you also know that it will not be happening anytime soon. Observational time constraints also often hinder collecting the information since the secondary tasks are only performed occasionally and later analysis must identify which are valid tasks that are interwove with some reasonable frequency and which were truly one-time tasks.

An extension to interweaving tasks and searching for information related to the current text is finding information related to a topic triggered by something in the current text, but not explicitly related to it. This can run from the "oh, I can use this for the usage report later today" to "Ginger and I were talking about this yesterday." There is no direct relationship which can be captured, thus reducing the usefulness of analysis methods in solving this problem. But it does result in task interweaving, because people will stop the primary task and pursue this new information for some time period.

## View of the real user

Too often the user groups are divided into two or perhaps three ill-defined groups: novice, intermediate, and expert. I claim they are often ill-defined because the categories are too soft. Consider the various ways one person can be an expert:

| | |
|---|---|
| Situation expert | The person understands the subject in great detail, but know nothing about to use the system to gain information about the situation. |
| System expert | A person who has spent several weeks doing a system test will have an immense knowledge of how the system works. However, they may have a very fuzzy understanding about how to apply the system to the situations of interest. For example, the tester can explain how to perform all the tasks in a marketing system, but has no idea of how to actually do the sales and marketing. |

| | |
|---|---|
| Partial subject or situation expert | The person is an expert on part of the situation or system, butnot on the aspect of interest. For example, a person is an expert at generating charts in Microsoft Excel, but has no idea how to insert conditional statements into a spreadsheet. These people show expert behavior in their area of expertise, but novice behavior outside it. If the overall system design spans areas in which these people will act as experts and novices, how should they be classified? As we'll see in the section on prior knowledge and mental models, this partial knowledge can prove detrimental. |

A prior knowledge effect on expertise can occur with seemingly minimal changes in the how the situation matches reality. For example, chess masters could recreate a real chess board setup after only a brief look, while with randomly placed pieces they did no better than chess novices. Novices, on the other hand, performed equally (poorly) for both real board set ups and randomly placed pieces (Chase & Simon, 1973; Frey & Adesman, 1976).

An interesting research finding, domain experts spend more time scanning and reading text than in formulating queries and general information search. They also spend more time initially analyzing a problem qualitatively, whereas novices plunge right in to setting goals and finding information. Experts also have stronger self-monitoring skills that allows them to spot an error or know when to double check information. (Andriole & Adelman, 1995) . They can also take material that lack coherence and still draw the proper conclusions from it (McNamera, 2001).. People that are referred to as experts seem to have an image of the answer and are guided by searching for answers which match that image. Obviously, within their area of expertise this a good thing. However, if the person is operating only slightly outside their domain of expertise, the mental image of the answer could be wrong, yet they will continue to search for it, assuming their knowledge applies (Marchionini, 1995). Novices, on the other hand, tend be 'sucked in' by the constraints of the information system. Because they have not yet acquired fixed working habits, novices may adapt practices that are easy rather than those that lead to good outcomes. And if the information system is not based on a sound work analysis, then what is easy is usually not what is good (Vicente, 1999a).

Although many research reports and design guidelines discuss expert and novice characteristics, a real problem with user descriptions based on a novice and expert dichotomy is the lack of clear middle ground. Novices remain so for only a short time and experts are only experts in limited areas. Also, most people do not advance to expert; rather than starting at novice and progressing to expert, people normally stop in the middle area and have no desire to learn more. People learn only enough to adequately address the situations in which they work. These make up the vast majority of users, who fall into the immediate category.

Fix, Wiedenbeck, and Scholtz (1993) found that all levels of users have at least the beginnings of the major abstract characteristics that are fully formed in expert views. Also, non-expert users often have an easier time bringing out their information requirements, since they have not internalized as much of their knowledge (Hale, Sharpe, & Haworth, 1996; McDermid, 1994; McNamara & Kintsch, 1996). Instead of getting hung up in definitions of novice and expert, the user analysis must strive to create a model of the goals and information needs of real users, who show a blend of the characteristics of both novices and experts (Santhanam & Wiedenbeck, 1993), not to create a model of the ideal expert (Johnson, 1994). Developing a model of real users requires collecting information from a full spectrum of users so the range of knowledge and detail requirements can be matched against their reading ability. Cooper's work on personas fits neatly into this requirement. By clearly defining all the user groups and figuring out how each one views the situation, it allows a clearer definition of understanding the person's goals and information needs without forcing that person into a predefined box.

## Understanding a situation

In any complex situation, a person receives too much information to process it individually. In response, people have developed cognitive processes that allow for promptly handling that information. We have each developed a personal information structure that includes all the filtering and organizing processes. The technical name for this processing method is a mental model (the literature also uses many other names such as schemas, scripts, or cognitive models). Based on that model, people quickly evaluate a situation by placing available information into waiting slots and form their goals and define their information needs (Kent, 1987). Obviously, if the mental model is wrong, the goals and information needs are wrong for the situation, although they may be right for the incorrect mental model. This section looks at how people go about setting goals and solving problems, and how mental models form the basis of understanding complex situations.

### *Recognition primed problem solving*

In complex situations and the information systems which support those situations, people must get enough information to quickly make some sort of appraisal. Users classify a situation and define what is relevant and what is irrelevant information. An obvious problem immediately come to mind: The wrong conceptual model means relevant information gets ignored as irrelevant, and vice versa. In general, the classical decision making model has been discredited as a model of how people actually make decisions. However, it is still prevalent in many texts and is often the only model presented. Thus, I'll briefly discuss it here in order provide a contrast with a more realistic model. A more realistic approach to how people deal with complex situations has been addressed by Klein's research in to recognition primed decision making.

### Classical decision making model

The classical decision making model attempts to quantitatively evaluate the optimal or best solution to the problem. In this model, the person is expected to evaluate all the various alternative solutions with respect to the factors that influence the solutions. The result is a matrix with factors across the top and solutions down the side. By assigning weighting functions to each factor, the optimal solution can be found by simply summing over the factors; the solution with the highest value is the optimal solution.

While looking nice from an empirical viewpoint, research has shown that people simply don't evaluate situations in this manner (Klein, 1999; Orasanu, Caldewood, & Zsambook, 1993). The obvious problems with the classical model are: (1) determining all the alternative solutions in advance, (2) determining the factors which influence them, and (3) setting the weighting factors. There is no method to prove all possible solutions or factors have been defined, or that the factor weights are realistic. Also, many people (the boss) respond to this type of analysis by adjusting the weighting factors until the totals on the matrix matches the desired decision; a decision not based on the analysis, but matching the one which the person wanted made regardless.

### Recognition primed model

Klein (1993, 1999) has been highly critical of the classical model and has advanced a recognition primed model which more closely fits real-world decision making. The recognition primed model assumes users perform assessments of situations based on experience and attempting to make satisfactory, rather than optimal, decisions. Field research indicates this fits people's real-life decisions much better than the classical model. A primary difference is that unlike the basic assumption of the classical model, decision making operates as a satisfying process rather than an optimal path process (Orasanu & Connolly, 1993; Simon, 1979). People often settle for less than optimal performance; instead of maximizing output, they economize on cognitive resource allocation and attempt to produce satisfactory output with minimal effort. Based on how they view the situation (what the next section will define as the active mental model) they rely on intuitive thinking, which rapidly leads to an answer but which follows a route that gives little information on how or why it was chosen (Rasmussen, 1986; Klein, 1999). This model fits the real-world process of a person quickly looking a situation over and almost immediately knowing what to do based on past experience.

In complex situations of interest to this book, the decision making may not proceed as quickly as the situations Klein studied, but a user will still tend to quickly develop a preferred solution or, at least, a partial solution. The recognition-primed decision making model brings out that, in contrast to the step-by-step analytical process of conventional task analysis, problem solving "rarely arises straightforwardly, but rather results from a long and recursive process with backtracking and erratic switching among the following activities: thinking about ideas, production, reorganization, modification, and evaluation" (Nanard & Nanard, 1995, p. 50).

## Mental model of application windows

The mental models people form are complex. When they don't know how something really works, people tend to formulate the simplest possible model. Like this example of how the windows work on a Macintosh computer.

Sit down at a Macintosh. Open two Excel spreadsheet files and a Word document file. Most users would guess that the windows are independent because they look independent. Thus, user mental model says that clicking on Spreadsheet 1 brings that window to the front. What really happens is that Spreadsheet 2 comes to the front. Not what the user expected based on the mental model.

As it turns out, Microsoft Excel's program model says that "you have these invisible sheets, one for each application, and the windows are 'glued' to those invisible sheets. When you bring Excel to the foreground, all other windows from Excel will move forward, too." Invisible sheets? What are the chances that the user mental model included the concept of invisible sheets? Probably about zero. So new users will be surprised by this behavior (adapted from Spolsky, 2003).

Klein also found people evaluate based on the order in which they receive information; accentuating the importance for the designer to present information in the order relevant to understanding the situation. Part of Wright's (1974) findings on time pressure and decision making support the idea that, when under time pressure, users want to quickly reach closure on the problem and will to jump to a conclusion before seeing all of the information. Often this results in their following impressionist or stereotypical actions and going with initial judgments, rather than evaluating the situation (Webster & Kruglanski, 1994).

As if handling complex situations is not complicated enough, most real-world situations either are not analytic in nature, thus not lending themselves to analytical task analysis, or situation complexity results in more data than can possibly be handled. Hollnagel (1993) adheres to Lindblom's model of "muddling thorough" and handling situations in a reactive manner with with decision making going through the stages of "(1) define the principle objective; (2) outline a few obvious alternatives; (3) select an alternative that is a reasonable compromise between means and values; and finally, (4) repeat the procedure if the result is unsatisfactory or if the situation changes too much" (p. 32).

### *Mental models*

As part of evaluating a situation, people use a mental model to fit the current situation into past experiences. A mental model (the literature also used terms such as cognitive model, cognitive schema, mental schema, or scripts), corresponds to the cognitive layout that a person uses to organize information in memory (Johnson-Larid, 1983). The mental model helps to make connections among disparate bits of information (Redish, 1994). In overly simplistic terms, a mental model is a template in the mind, built on previous experience, that contains a collection of known information and relationships for a particular class of situation. All new information gets fitted into this template. Readers interpret information by applying their own knowledge and expectations to the template. People use a mental model as a basis for understanding the situation and making predictions about future events. The classical textbook example is the difference in the mental model between a fast food restaurant and a restaurant with table service. If you walked into a fast food restaurant and were greeted by a host wearing a tux, you would be confused as this does not fit the model. Likewise, you would be confused if you entered an expensive French restaurant realized you were expected to place the order at the counter.

Mental models work because they provide a preset structure on which to interpret observations of highly complex situations. They contain both the structural patterns that define information relationships and information importance. The mental model allows the user to apply it to various situations in order to make inferences and fill in missing information (Fawcett, Ferdinand, & Rockley, 1991; Smith & Goodman, 1984). A mental model is a stereotype of the situation, not a detailed description. Thus, your fast food mental model works at for all the various chains although they each handle food ordering in slightly different ways. However, it provides people with a structure into which to place and interpret specific details. By applying a mental model to a situation, a person can quickly place everything within a context and make the proper response.

According to Anderson and Pearson (1984), mental models serve as networks linked by factors that imply order, such as chronology, function, topics, and so forth. The user's mental model strongly influences the view of the world and how events/facts are interpreted. If the design and presentation of the information matches the mental model, it's easy to process. Comprehension of information requires the presented information mesh with and complete the mental picture of the active mental model currently being used to represent information and its relationships. If the reader has to work to match the information to the mental model, it takes more time and is much harder to understand. Design failures arise when a design is not sensitive to or fails to match the users' mental models (Gribbons, 1991). This view of mental model construction treats information coherence as a positive factor and cognitive overload as a negative factor (Thuring, Hannemann, & Haake, 1995).

### *Mental models applied to complex situations*

The basic framework for acquiring information in complex situations is human-centered in that the user defines the task, controls the interaction with the system, examines and extracts relevant information, assesses the progress toward an acceptable understanding, and determines when adequate information has been collected to meet the current goal.

The high level structure of the user goals and the information relationships corresponds to a mental model. The structure of the goals (i.e., order of addressing goals, which sub-goals go with higher goals) form a model of how the user views the situation. The information relationships show how factors within goals are related. Structural patterns inherent in mental models are essential to human knowledge representation; the identification and representation must carry over to the interaction with computer systems (Treu, 1992). Each time users encounters a similar situation, they initially set similar goals and expect similar information relationships. They use their mental model of the situation to define goals and relationships and to decide on which information they need to achieve those goals. Thus, the user should receive information about the situation in a manner that matches the mental model.

### *Capturing user mental models*

With good mental models of situation behavior, users have knowledge for "the dynamic direction of attention to critical cues and for expectations regarding future states of the environment" (Endsley, 1995, p. 44). The difficulty with matching mental models to user goals and information needs arises because each person tends to have some unique features within their mental models, experiences, abilities, and preferences (Marchionini, 1995).

To support system design we do not need detailed process models of the mental activities users. Instead, the design must be based on higher level models of the structures of effective mental processes which are used and their characteristics with respect to human limitations and preferences, so people can adapt individually and develop effective strategies to handle the situation. Rather than descriptions of the course and content of actual mental processes, design analysis needs to define descriptions of the structure of possible and effective mental processes. (Rasmussen, 1981, p. 242 cited in Vicente, 1999a, p. 215). The analysis for the design of complex situations should work not uncover specific user mental model of the system but the entire understanding of the situation in both the specific instance and the stereotypically model of the situation.

Both Rasmussen (1986) and Terveen, Selfridge, and Long (1995) discuss how if the designer had a consistent method of matching the user mental model to the system model, design would be much easier. However since each user is different, user analysis which aims to construct only one representation runs a high risk of being minimally applicable to a substantial number of users. Unfortunately, such a method does not exist because of the wide variation in users and situations. There simply is not a single mental model used by all users when they consider a complex situation. Each person has a slightly different mental model: Two people can observe the same information for the same reason and, based on their mental model, conceptually organize the information differently, select different information for deeper analysis, and draw different conclusions. For that matter, even in a situation as simple as ATM use, users can have many mental models, some of which extreme misconceptions of how the system works. Understanding that different mental models, many with flaws, exist and will influence the user's interaction with and understanding of the information and its relationships is a critical part developing information for complex systems. Thus, user analysis must uncover the variations and differences in the mental models within the intended audiences, rather than searching only for similarities.

## Mental models change over time

Mental models affect one's expectations for what happens when performing a certain action. Consider using Amazon's search engine to look up books title. Because it is flexible and good, and because many people use it often, it forms their mental model of how book searching is done with a computer.

Now have that same person search my university's online library catalog. A fundamental difference is that the university catalog doesn't allow free text searching; a search must start with the first word of a book title. For someone not spoiled by Amazon that may not be a problem. In fact, until Amazon, most people only searched for books with library catalogs and expected they had to know the first words of the title. But now it doesn't conform to current mental model of searching for books should work.

Blandford and Duke (1997) call for theory-based analysis to help derive the deeper issues. Their theory-based approach provides a basis for an explanation of the observed user actions and for generalization based on specific instances. Theory-based analysis helps because capturing mental models is problematic since they tend to operate below the conscious mind. People great trouble explaining what they do or how they do it. Thus, we must both ask them what they do and also watch them do it. Unfortunately, researchers frequently find the two have little in common. Part of the reason is our brain is hardwired to quickly classify situations and disregard the normal every day aspects. People rarely consciously consider the normal, everyday things around them, they focus on the unusual or out of place items. Not a bad evolutionary adaptation, but one which complicates design analysis by making uncovering a user's mental model extremely difficult.

Two questions facing every designer is, "What is the best way to organize the information for the audience?" and "what information needs to be presented and how?" Hackos (2002) very neatly sums up the reason for looking at mental models and connects it with the primary points of this book:

> To know how to present information effectively to our users, we must concentrate our investigations on understanding existing conceptual [mental] models, especially in terms of the context in which information is understood. We need a comprehensive understanding of what users need to know, when they need to know it, and in what ways they attempt to interpret information and turn it into analysis and action.

Besides providing information fitting the user goals and information needs, final design must provide the information in a way that corresponds to how a user expects the information to appear (matches the user mental model). A big problem to answering any question with *the best way* arises because each user's goals tend to redefine what "best way" means. As was shown in chapter 2, complex situations have no one best way. The appendix describes a graphical method of capturing how various users structure information. After a user analysis provides this information, the designer can use it to gain insight into how the user mentally structures and thinks about a situation. A user which is never static, but always changing.

> A user's model is not methodically thought out, but instead grows rather organically and spontaneously from the process of interactions with the user interface. The user slowly develops a picture of what is in the system, what it is capable of doing, and how it responds to various actions. (Constantine & Lockwood, p. 42)

## Defining the goals

Design for simple situations concentrates on giving access to single pieces of data, rather than supporting the ability to merge individual pieces of data into information and to get a complete picture of the situation. Whereas simple look-up works for some cases, in general, a user's goal is much more complex and corresponds to their mental model.

People have a huge collection of mental models and activate a model that they consider appropriate for the current situation. Mental model activation occurs rapidly and at a subconscious level. Thus, a person will always have an active mental model to interpret a situation. Ensuring the proper mental model gets activated means that the design must immediately provide a view of the information which fits with the user's expectation. A design always invokes a response; the designer must ensure it's the proper response.

Since the active mental model determines information interpretation, designs that initially invoke the wrong mental model pose trouble from the start. A user goal is set based on current mental model; an incorrect mental model can cause goal to be set which is incorrect for the situation (but correct for the active mental model). With the wrong mental model, misinterpretation of information becomes almost certain. For example, Duin (1991) found that poorly organized documentation failed to activate the appropriate mental model mind and cause the user trouble in understanding the documentation and relating it to the system. Thus, a major responsibility for the designer is to ensure that the proper mental model gets activated because all information gets interpreted with respect to it, and people resist changing to a new mental model.

Systems for complex information require a through user analysis and continual usability testing to ensure the design contains all the features needed to invoke the proper response and that it is laid out in the manner which users expect. The preceding sentence is a restatement of the design guideline that a properly designed interface transforms a system model into a user's model and does not force the user to think in terms of the system model. When the information is presented in a manner to match the user's mental model, it enhances effective system use and helps ensure the user gets a clear picture of the situation and sets the proper goals. The interface becomes an important part of the understanding the situation as the user assimilates the information through all the means provided in the interface (Treu, 1990).

Note that the concern here is with the design and user interaction. The structure and implementation of the underlying system design is irrelevant to this discussion, except as it supports the user interaction. The interface must reflect the user's mental model so that it fits into the user's world and not the programmer's world. The user's mental model does not reflect the underlying system model of the program's internals; rather it must be reflected in how the interface appears to work.

## *Expectancy bias*

Expectancy bias is the problem of seeing what a person expects to see and ignoring other information. Based on past experience, in a particular situation, a person expects certain things to be true or certain actions to occur. The expectancy bias comes into place because in most instances, the person will see these things, even when another person without the bias does not see them. Likewise, they may ignore blantant aspects of a situation because they do not expect to see them. Rather than being highly overt, many times this is subtle interpretation. For instance, if a person expects an establishment be dirty, they will notice any and all dirt on the floor. But in an establishment the expect to be clean, they will ignore the dirt even it exceeds the amount in the place expected to be dirty. The same type of bias arises in discussion of how more disgusting and rowdy one sports teams fans are than the person's favorite team fans. A writing example: If a writer comes at a situation having already decided to write a reference manual, all information gets interpreted with respect to how it applies to a reference manual. Factors indicating an online system would be better get ignored or discounted. The expectancy bias lead to seeing what is needed for a reference manual and ignoring all else.

Expectancy bias is not all bad. It is how we can operate in a normal world without suffering information overload from all the actions occurring around us. In most instances, what we expect to see is what is actually there so the expectancy bias saves cognitive resources by simplifying how much work is required to interpret the world. However, when it starts to affect getting a clear understanding of a complex situation, then problems occur. Unfortunately, expectancy bias also tends build up over time and is only seen in hindsight (Klein, 1988). Many post-failure reports discuss how the problem should have been obvious, but the people kept ignoring them or dismissing them as not significant.

Information presentation has major effects on people's decisions; depending on the presentation, they may actually make opposite choices and believe them to be best (Tversky & Kahneman, 1981; Johnson, Payne, & Bettman, 1988). Even with complete information, different presentations cause different approaches to the solution (Elam & Mead, 1990). When people look at information, expectancy bias causes them to see the expected answer. Presenting the information in different order caused the information to be incorrectly interpreted with respect to the mental model.

## *Effort required to achieve the goal*

Databases cannot simply connect to screens. The display format can actually change the way users view the information. Slovic (1972) found that users tend to change their assessment strategy to fit to the presentation method, rather than transform the information to fit a better assessment strategy. The amount of perceived cognitive effort seems to drive how different displays affect assessment strategies. "Since different display formats affect the effort required by various strategies, decision-makers may react to changes in display format by adapting strategies which minimize effort" (Johnson, Payne, & Bettman, 1988, p. 2). Thus, display changes that make it easier to process information also increase the information's impact upon the decision-making process.

The failure to change assessment strategy arises because people are very limited in their ability to handle multiple mental tasks (Wickens, 1992). How well the task fits into a mental schema directly affects perceived complexity. Kieras and Polson (1985) showed that task complexity comes from the user's knowledge of the task. In other words, the difficulty in learning the knowledge necessary for the task is often independent of the difficulty of performing the actual task.

Also, as the decision task increases in complexity, rather than increase the complexity of the decision making strategy, relatively simple, error prone heuristics are used (Fennema & Kleinmuntz, 1995). The heuristic used depends heavily on how the information presentation and the anticipated effort of extracting the desired information. The decision makers mentally create a goal frame around the decision and, as a result of expectancy bias, give preference to information which supports the expected results. As the amount of information increases, the users increasingly ignore information incompatible with what they expect (Ganzach & Schul, 1995). Although this seems to support a highly compact information presen-

## Define success in context

A problem with defining success is that most metrics are based on usability tests that were developed for applications used in simple tasks and limited work contexts. In these situations, it makes sense to observe and measure discrete tasks and clear success metrics (time-to-completion, etc.)

When it comes to consumer-oriented Web sites, though, things get fuzzy. "Number of clicks" or "time to completion" aren't very worthwhile measurements, unless the action is a extremely highly specific task such as placing on order (Ghedira et al., 2002). But placing an order bears little relationship to the environment for a more experiential, wandering task like researching products. Here the user is working in 10-15 minute (or longer) intervals, at work and may have to handle many distractions in the process.

Although a performance test can say that a user can get to a product in X clicks or X seconds, that's not really applicable to complex situations. What's more important, and well-nigh impossible to measure is if the system helps the user understand the situation in an efficient or effective manner. The use may have gained an understanding, but as it efficient? And what does "efficient" mean in this context? You must know the user's goals and see how efficiently the site provides the means to address those goals and lead the user to an answer of the ultimate goal.

tation, Rubens and Rubens (1988) found that making the information too compact or concise hindered performance. Rather, information usage and how it relates to understanding the situation must occupy a prominent position in design considerations and the information structure must enhance its usability.

People have an impetus to move quickly to closure and claim to understand the situation by using strategies based on a trade-off between amount of cognitive effort and information accuracy. However, the effort used for strategy selection is not actual effort, but anticipated effort (how hard does it *seem* to be if this strategy is used). Unfortunately, "individuals are better at effort anticipation than accuracy anticipation" (Fennema & Kleinmuntz, 1995, p. 23). To complicate the situation, as situation complexity increases, accurate judgment of both anticipation and accuracy decrease. When users meet up with complex situations in which the design does not effectively assist in solving, contrary to the desired result, the users' ability to understand tends to decrease and they resort to simple heuristic rules. Unfortunately, decision based on this type of understanding often leads to non-optimal solutions and may lead to incorrect solutions. To further complicate the design, as a mental planning strategy, rather than scheduling tasks in proportion to their significance and

## Problems caused by an inappropriate mental model

Essentially all my students are experienced with Microsoft Word. Of course, whether they have a clear mental model of how a word processor differs from a typewriter covers the entire gamut. For example, many use the space bar to center text, rather than using tabs or the format-paragraph command.

On the other hand, very few have experience with Adobe PageMaker or any familiarity with old style paste-up work. When I first introduce them to PageMaker, they tend to encounter many errors based on using a word processor mental model. Performing a relatively simple task, such as designing a one page flyer, causes them no end of trouble because everything get put into one text block.

The problem here is that they are working with a word processor mental model which says, among other things:

- Text must start at the top of a page and got to the bottom, even if just by pressing the return key.
- The entire block of text does not move around the page when you click and drag on it, unless it was highlighted.
- Text or graphics are either on the page or deleted. There is no such thing as the desktop for holding text or graphics.

Until they grasp that PageMaker uses a paste-up model, rather than a glorified typewriter model, they have trouble. Until they make that mental shift, the fact that individual text or graphic blocks can be individually placed and manipulated, make no sense to them. After they formulate a "PageMaker" mental model, they become more productive as they work within the proper model versus thinking they are using a word processor.

cognitive demands, people tend to use equal-scheduling which assigns each task the same amount of mental effort. The result is a mismatch of attention resources to demand as the needs of some tasks cannot be filled while others receive an excess (Langholtz, Gettys, & Foote, 1995).

In Timmerman and Vlek's (1994) study of decision support system effectiveness, they found that, while effective for moderately complex problems, for highly complex problems, the decision support became less time efficient. However, user perception enters into making use of they system and how complicated the system is to use. Although, for highly complex problems, computer assistance was less efficient, users had decreased perceptions of task complexity. Thus, the designer must make a trade-off between time efficiency and user perception of task complexity. In many cases, especially for non-time critical tasks, user perception should far outweigh other concerns.

In a study of business managers, besides failing to attempt to obtain exact results, they exhibited decision making that showed both risk-taking and avoiding behavior. Rather than consider each situation individually, the managers tended to base decisions on recent events with the greatest risk taking appearing after a gain. However, overall, the results showed an underlying tendency toward risk avoidance (Sullivan & Kida, 1995). Besides avoiding risk, managers also failed to anticipate losses or exploit gains, but instead adapted a less-than-optimal strategy, consuming resources as if neither a loss nor gain would occur (Langholtz, Gettys, & Foote, 1995). Langholtz et al. attributed part of the less-than-optimal behavior to an equal-scheduling tendency for resource-allocation problems: rather than partitioning resources by their importance, managers tended to divide them equally.

The risk-taking and avoidance behavior becomes even more pronounced when the decision makers are under pressure. Wright (1974) found that time pressure and distractions had major influences on decision making. When under pressure, users adapted simpler strategies that included ignoring less important data or focusing on only certain regions of the data. With time pressure and distractions a normal day occurrence, failure to account for these in the design can lead to erroneous decisions. Although the normal routine of an office provides numerous distractions, even distractions as simple as having to jump between multiple windows, or refer to manuals or online help may influence decision strategies. Currently lacking, further research into examining distractions within the information presentation itself seems worthwhile.

## *Goal and situation reality mismatch*

For good or bad, once activated, all new information gets interpreted with respect to that mental model. Consequently, inadequate designs run a risk of improper mental model activation and the subsequent risk of incorrect action. Using an incorrect mental model compromises how people understand a situation. They believe they understand the information, but do not have the interrelationships correct and, thus, do not correctly comprehend the situation. The design problem is that once they start using a mental model, people have a hard time shifting to another one. The effect can be so strong that people refuse to consider other interpretations (Spyridakis & Wenger, 1992) and will force information to fit their preconceived expectations. Once people activate an incorrect mental model, they do not switch to a different one until problems with data interpretation reach a highly critical stage major logical disconnects between reality and their mental model.

After-the-fact analysis of many errors in human performance reveal that the users based their decisions on the wrong mental model, and consequently, performed correct actions for the schema which were incorrect for the actual situation (Wickens, 1992). Also, post-failure analysis of design problem often reveal the user's mental model envisioned by the designer departed from the actual user's mental model; a difference which often hinged on the social context. The mental model made sense in the designer's social context; however, it failed to make sense in the users' social context. All too often, the social context was totally ignored during the initial analysis phase, leaving the designers with only a technological context in which to work and, consequently, without a complete picture of the situation.

## Interpreting information

Most systems, especially ones that address complex situations, have a high information content, but, poor design often results in low information communication because it fails to conform to the user's mental model of the situation (ultimately resulting in the failure of the system). Although the information may exist within the system, the user is unable to interpret it and apply it toward achieving a goal. From the psychological standpoint, the information is not disseminated effectively.

Undisputedly, simply having volumes of data/information available has proven insufficient (Mirel, 1996). In the end, as the user cannot find or develop the necessary information relationships, the online information fails because it fails to anticipate the users' real-world needs.

Providing information which users can properly interpret with respect to their goals, means accepting the complexity of the real world and try to work with it (Woods & Roth, 1988; Rasmussen, 1986) rather than simplifying the situation as is common in many documentation and software engineering practices. This simplification was mentioned in earlier chapters as a problem of taking complex situation and trying to define it as a set of simple tasks that can be performed in a linear sequence. Instead, always design to address a real-world situation that a user may encounter, or as Beyer and Holtzblatt (1998) put it, "anyone's real work practice is intricate and complex" (p. 3).

In contrast to a poorly designed system, the goal of a well designed system is to provide the user with both the proper information and the proper amount of information. The design of a system must strive to be "so proximal that the user's flow of creative thinking is not interrupted" (Basden, Brown, & Tetlow, 1996, p. 157). Any gap between the user's model and the system model causes a deficiency

## Table of numbers with good and bad arrangement

Here are three different arrangements of hotel information. Assume the first two contain all the corporate-approved hotels in Charleston and Columbia, so they could be much longer. Obviously, everyone would pick the first list over the second because it is so much easier to read. Yet, both lists contain the exact same information; the difference is all in the presentation.

If the lists were long or you didn't know where any of them were located (it's your first trip to Charleston), the a long list of all possible Charleston hotels is not too useful. You might pick a major chain, only to find its on the opposite side of town from your business appointments. Instead, to really support achieving your goals, you want the list shortened to only those hotels close to where you are working. In this case, would you prefer list 1, or list 3, with only 2 hotels listed?

List 1—all corporate approved hotels in Charleston and Columbia

| City | Hotel | Phone | Rates |
|---|---|---|---|
| Charleston | Best Western | 555-0961 | $53 |
| Charleston | Days Inn | 555-1393 | $48 |
| Charleston | Holiday Inn | 555-2395 | $63 |
| Columbia | Best Western | 555-9400 | $59 |
| Columbia | Carolina Inn | 555-8282 | 7$7 |

List 2—all corporate approved hotels in Charleston and Columbia

| | |
|---|---|
| Charleston: Best Western<br>555-0961 $53 | Columbia: Best Western<br>555-9400 $59 |
| Charleston: Days Inn<br>555-1393 $48 | Columbia: Carolina Inn<br>555-8282 $77 |
| Charleston: Holiday Inn<br>555-2395 $63 | |

List 3—only hotels close to your appointment

| City | Hotel | Phone | Rates |
|---|---|---|---|
| Charleston | Best Western | 555-0961 | $53 |
| Charleston | Days Inn | 555-1393 | $48 |

in both usability and information communication (O'Malley, 1986). Part of the reason ensuring a user looks at both the proper information and the proper amount of information provides a major challenge is because a user rarely know exactly what to ask for. Instead, the designer must be concerned about what information is required to allow the user to form the proper relationships and must be concerned that information is salient.

People have a very hard time working with information, often working very close to information overload. Also, they often don't know what information even exists, are easily fatigued, and often simply look to verify the solution they already want. Even in the best case when the person generally knows the contents of all the documents (for instance, an executive reviewing a set of monthly reports), the task of mentally coordinating and relating the information contained in all the reports increases at a highly nonlinear rate with the number of reports and report complexity (Carlson, Wenger, & Sullivan, 1993; Mirel, 1998).

### *Salient information to define relationships*

A real problem in interpreting information is to not be distracted or mislead with information which looks good but is not relevant to the current needs. The increased use of technology to provide information makes it easy to provide too much of the wrong information. Forming the proper relationships between information is easiest when the most salient information occupies the most prominent locations.

Unfortunately, too often system design often ignores the problem of how to effectively handle unstructured data, or ensuring information receives a focus proportional its salience. Instead, everything gets treated equal, with the user having the responsibility of integrating the data into a coherent collection of information, a. line of reasoning detrimental to effective understanding (Wickens, 1992).

The user and task analysis have define the important information, the proper order of presentation, and the relationships which should be formed. The designer then needs to ensure the system presents this information to the user in a manner corresponding to that analysis. The salient information must be both easy to see and easy to formulate into the proper relationships.

Presentation with proper salience is vital for unsolicited information (Owen, 1986). As was mentioned earlier in this chapter when discussing mental models, most people respond by ignoring the new information By not seeing how it fits into the current mental model, it gets discarded as irrelevant. Of course, when the information is relevant, the person risks interpreting the rest of the information incorrectly.

### Information disregarded because its in the wrong place

I recently decides to use the online system to renew a prescription. As you might expect, the web site guided me through several steps. First I entered the prescription number, my name and address, when I wanted to pick it up. After I was done, I noticed that the designers are provided a visual of my location in the process. There on the left hand side of the screen were all five steps with checkmarks beside them. Because I was done, I can only assume they added a checkmark at each step in the process.

But I didn't notice that indicator until I had entered all the information as was looking at the final screen that told me the order had been entered. Only then did I shift my focus from the right hand work area to the rest of the screen. Those steps and checkmarks were placed in the left hand area of a web page, the area normally occupied by the navigation menu. As I didn't intend to navigate anywhere, I disregarded all information in that area of the screen. If early in the process, I had wanted to know how many more steps I had to complete, I still probably would not have seen those. My mental model told me anything along the left was navigation information, not 'placing my prescription order' information, and thus, I could safely ignore it.

## Close enough is good enough

People often need to quickly interpret rapidly changing information (Schamber, 1995). In the process, they make decisions about how to achieve a goal as quickly as possible, and abandon the information analysis/interpretation as soon as they feel they have adequate information to achieve the goal.

A common misconception in information design is that the person wants to find the best or perfect answer. In reality, research has found that close is good enough. It seems that people know their choices is not perfect, but is good enough. Rather than attempting to find "the answer," people are content to make a decision which moves them in the proper direction and continue to make follow-on decisions which function as adjustments to the initial decision. Information design must focus on getting the user properly close enough and the information must be structured to enhance its usability.

People use an informational system to gain an understanding, accepting that the system often contains incomplete information. Accepting as a given that the information is incomplete, people rarely strive for an optimal answer, but rather are content with a sufficient answer, making a trade-off between the available time and available information (Klein, 1988; Orasanu & Connolly, 1993). As a result, people often make choices based on using the first alternative (Klein, 1999; Rouse & Valusek, 1993).

In addition, in complex situations "best" has little meaning and people must have the ability to solve the problem in their own individual ways, especially since many decisions actually have a "fuzzy" success criteria (Hallgren, 1997). Unfortunately, the idea of a best answer often carries over to the usability testing of a system. Larichev et al. (1995) claim that, since people are not concerned with exact answers, it makes no sense to measure systems based on "exactness of results" (p. 18). Yet, design guidelines and usability test criteria commonly use exactness of answers as their definition of effectiveness. In complex systems, rather than ensuring the user has found the best answer, user tests should ensure that the user's answer is acceptable and would result in a proper outcome, even if it is not an optimal strategy.

## Information overload and information dumping

As the amount of information increases in complexity, contrary to intuition, people do not increase the complexity of their analysis strategy. In fact, it seems as if "decision makers will not be able to correctly anticipate the simultaneous influence of more than one task feature" (Fennema & Kleinmuntz, 1995, p. 23). In complex situations, rather than trying to integrate multiple informational elements, users tend to make choices based on methods that are error prone, but easy to use. They avoid information overload by shedding tasks and trying to mentally simplify the analysis.

For example, consider if a person had many different databases and other programs to use in obtaining information required to understand a situation. When the number becomes too big, ironically, only a few are used. The problem from the quality point of view becomes ensuring the proper few were used. Unfortunately, chances are good the main factors driving the choice is familiarity and desire for closure, not appropriateness. People tend to always use the ones they have used in the past; if other data sources have better information, it will never be used.

Cognitive overload occurs when the person has exhausted all the available cognitive resources. Because cognitive resources are very limited, unfortunately, cognitive overload can occur easily. When it occurs, several things happen:

- Error rates dramatically increase. Information may be accepted without evaluating its quality. The fails that occur when playing the higher levels of Tetris occur because of cognitive overload.
- Tasks are shed. Rather than trying to mentally juggle multiple tasks, the user sheds one or more. A common psychology test is to have the person rhythmically tap a key while performing a mental task. When overload occurs, the tapping stops. (Try multiplying three digit numbers while tapping your toe.)
- Information is disregarded. Instead of working to incorporate information into the current mental model, all information is ignored. This is basically a variation of task shedding.

Information must be filtered to prevent information overload (Bowie, 1996). Trying to adjust for cognitive overload within the design process is complicated by the constantly changing load levels as the users' experience levels change. With increasing experience, the level of cognitive processing drops. Experts handle larger amounts of information as single chunks, thus handling more information before cognitive overload occurs. The more complex the task and less the experience, the higher the task load (Neerincx & Griffioen, 1996). Unfortunately, trying to handle the problem of different knowledge levels often results in tradeoff that can compromise the effectiveness of the system for all levels of users (Herbig & Kramer, 1992).

### *Effect of prior situation knowledge*

A major factor affecting how people interpret information is their prior knowledge and experience, which often led to the original creation of a mental model. In general, researchers have noted that prior knowledge can affect comprehension either for the good or bad by activating a mental model that may or may not be correct (Spyridakis & Wenger, 1992; Terwilliger & Polson, 1997). An incorrect mental model, even one that is close enough most of the time, can cause interpretation problems once a situations occurs which is not quite close enough.

Intuitively, a designer would expect users with extensive prior knowledge to perform better at interpreting the information and understanding the situation. As Marchionini (1995) states "experience with particular settings, domains, and systems generally allow more comprehensive and accurate mental models and thus more facility with these models" (p. 33). With a solid background that includes past experience with similar situations, the user can make connections between disparate pieces of information and mentally fill-in missing pieces. As such, providing too much of this background information to the user with extensive prior knowledge bogs down the presentation with information the person does not need. Of course, other users lacking that prior knowledge and experience will need this information; a strong reason for dynamic information presentation.

With prior knowledge issues influencing design, it should be obvious that it complicates design that crossing different groups of users. Each group has different prior knowledge and each user group puts a different twist to their interpretation which must be accounted for (Smart, 1993). The user analysis must examine what prior knowledge the user group possesses and consider how to prevent inappropriate mental model activation. This means attempting uncover key words, concepts and proper terminology which will activate the proper mental model (Duin, 1991). Although not often explicitly brought out in the analysis literature, a fundamental objective of the user analysis is to uncover the mental model held by the user, implement the critical cues for mental model activation, and uncover methods of leveraging that model in the design.

Understanding the different needs can be difficult for an information designer, partially because the system's user base and the information designer may both have deep understandings of the context, but from very different viewpoints. In other words, the prior knowledge is radically different. A fundamental design problem can be getting the designers to see the problem from the viewpoints of all the user groups, an issue (lament?) that repeatedly appears in the usability literature as the problem of programmers not understanding users. Interestingly, it can be easier to for a designer starting with minimal knowledge rather than a deep, but wrong viewpoint knowledge, to understand the reader's viewpoint. Although, during the course of the design methodology, the designer must still acquire a deep knowledge of the situation context.

However, in some situations, subjects with no prior knowledge performed better than subjects with inappropriate prior knowledge (Lipson cited in Spyridakis & Wenger, 1992). One explanation for this non-intuitive result is that prior knowledge overrode new information which was not consistent with the prior knowledge. In other words, within a situation, people expect certain values for certain

## Superstitious behavior in software programs

Many people exhibit superstitious behavior when using a program. Because they don't understand how the program really operates, they have found, by trial and error, some method which produces the proper result. This method often includes steps that make no sense with respect to accomplishing the actual goal.

Psychologically, what has happened is the person searched for and found a sequence of actions that performed the desired task. Lacking knowledge to evaluate the sequence, the person performs the entire sequence every time they want to accomplish the task. Interestingly, pointing out the problem is likely to generate a response of "this works, I don't want to change."

information. If that expectation is wrong and the information has a different value, people tend to disregard the real value and mentally substitute the expected value. Then, rather figure out why the information is not what they expect, people ignore information or twist its interpretation to fit how they want to see it.

## *Effect of time and stress*

Many factors affect how people's perceive and interpret the information to gain an understanding of the situational context, with stress and time being the major factors. A pertinent point about stress and time is that they are often data-independent. Instead, they arise from the social aspects surrounding the situation. As both factors derive from social context and are highly unpredictable, they get ignored in most design analysis, yet they play a major part in determining whether readers consider the information system highly usable, mediocre, or unusable. Development of highly effective information systems must consider the way real people behavior in the real-world situation. Ignoring behaviors such as time constraints or situation-induced stress because they seem too difficult to capture or categorize impends the development of an a high-quality model on which to base the design (Allen, 1996).

The time and stress pressure may be inherent in the situation itself, as in a fast paced work environment. Stress levels may also move from minimal to high depending upon external factors. Ensuring the people have easy access to the information needed to maintain a firm grip on the contextual aspects of the situation mitigates these factors. Unfortunately, the stress effects do not show a consistent increase as stress increases, making it hard to allow for. Often stress factors play a trivial role in decision making until certain cognitive load levels are reached and then they rapidly assert a detrimental influence as shown in Figure 5.1 (Wickens, 1992).

The method of dumping information on users and hoping they can sort it out (the basic idea behind current Internet search engines) fails because it does not contribute to controlling the information presentation. Current trends of presenting information online fails to account for the dynamic, interactive potential of the online environment (Heba, 1997). In a time pressured environment, readers need to quickly obtain information relevant to the situation.

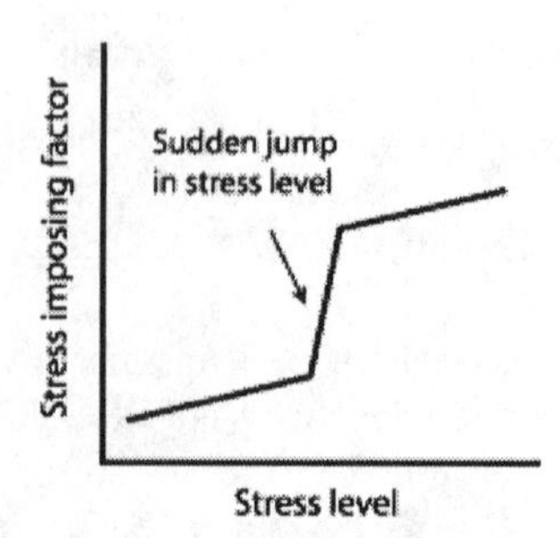


FIG. 5.1. Graph of stress jumping as a discontinuous function

Time and stress are hard to adequately account for in a design because during usability testing the user feels little of their real-world task-related stress and will plow through a poor system to reach the answer. Also, users may not either know what information may relate to the situation or may not consider all relevant information. But in a high-pressure job, the system must provide the answer without the "plow through" aspects. In high stress situations, people tend to get cognitive tunnel vision and only sample what they perceive as the dominant information sources. As a result, users often proceed along a path that often consists of using the first alternative (Klein 1999; Rouse & Valusek, 1993). It rests upon the information designer to translate the user analysis into an effective implementation that helps ensure the first alternative is a sufficient solution.

With the wrong mental model, they might sample the wrong information or reach premature closure (Endsley, 1995). Likewise, under time pressure, people's desire for closure leads them to rapidly make a decision (Klein, 1988; Wright, 1974). Under time pressure, rapid closure results in following impressionist or stereotypical actions and going with initial judgments (Endsley, 1995; Webster & Kruglanski, 1994). The problem with stress and time comes about because, when people look at information, expectancy bias causes them to see the expected answer (Klein, 1988). Rather than searching for a good solution, they look for information which verifies they answer they want and ignore information that might refute it. Good design of complex information must confront a user's verification and ensure the salience of the information which can refute the initial (most likely) choice.

## Conclusion

Carroll (1995a) states how the ideas for handling the people aspects of information need to be much more flexible than many current methodologies.

> We have little prospect of developing *final* answers to questions about the nature of human activity—certainly not a the level of detail that would provide specific guidance to designers. Our best course is to develop rich and flexible methods and concepts, to directly incorporate descriptions of potential users and uses they might make of an envisioned computer systems into the design reason for that system (p. 2).

People come to understand a complex situation based on how they interpret the information around them. Incorrect or incomplete information can lead to incomplete or invalid decisions. In a typical time-pressured environment, readers need to quickly obtain information relevant to the situation. Otherwise, the humans' desire for problem closure (Webster & Kruglanski, 1994) and trait of accepting the first logical answer (Klein, 1999) leads to potential problems. The user picks a workable solution, but risks it being highly sub-optimal, or, by disregarding important factors, incorrect. Building on the way in which people rapidly assess situations and make decisions requires considering information requirements in the light of models like Klein's recognition-primed model and considering how the context drives the decision-making process.

Effective design of systems which address complex situations requires an understanding of the user's mental model. It must clearly relate with the user's world view from the system's point of view. Designs with unclear models or designs which deviate from matching the user's model impair the user by their poor usability. Murray (1995) lays out the issue by saying:

> If the text is ill-formed (e.g., disorganized), it is more difficult to recreate our schemata [mental model]. Readers rely on text organization to help them access the information. They expect and use organizers such as headings and overviews; they expect and use cohesive devices. They expect a text that is coherent within their cultural framework. Western writing expects the writer to provide these cues for the reader. But, in hypertext, we expect the reader to do this. Moreover, experts in the knowledge domain of the text can crate representation when the text is organizes in an unpredictable manner. Novices in the knowledge domain can not. Therefore, it is even more important for the information to be organized in an expected manner since it is most often novices who are seeking the new information (p. 140).

Effective design which minimizes cognitive load does not eliminate the massive piles of information that reside within the system, but rather provides an effective means of communicating that information. The approach of simply placing information before readers has been tried and found wanting. Instead, the system must provide different perspectives on the information for supporting different users and the ways one user might need to view it (Benyon, 1998). The information must fit user goals, their needs and current state of knowledge rather than the systems (Belkin, 1980).

## Long example—Pricing of wholesale products

[Disclaimer: The business logic of this example is flawed because of the changes I made as part of hiding the company's identity. However, this is an example about the people issues in communicating complex information, not a description of how a sales force actually sets prices.]

Essentially every complex system works with many more elements that a simple set of figure. Too often, the system design and information presentation focuses on presenting single figures. Although those figures are required, they must be placed in the proper context. Unless the person can understand how they fit into the overall process, they cannot fully understand how to make decisions which result in a new or modified set of figures.

In this example, assume a wholesale product pricing and invoicing system. (In this example, customers are retail stores, not the general public.) Orders are faxed or phoned to the company, workers fill the order and takes it to the pricing station. At the pricing station, a person scans the products, prints and inserts an invoice, and seals the box for delivery to a customer. Based volume, product mix, and other marketing-based factors, each customer has a separate price schedule. In other words, two different customers may pay a different price for product X.

Each sales person has a laptop that is synced daily with current client information. It also contains custom software for price and order inquiry which is supposed to be used to support setting prices.

The prices are very dynamic with salesman constantly changing prices because of factors such as manufacturer price changes, promotions or the need to undercut the competition across the street. Especially for issues like matching the competition, the price changes must be made right away and may even be retroactive (what was bought two days ago gets this new price too). The system must provide information for the sales people so they determine the proper price. Previously, they were going too low and costing the company money, or the customer was buying from a competitor because the price was too high.

In any sales job, the primary goal is to make sales to the customer. In this case, it often involves constant renegotiation of individual product prices to meet local business conditions.

It is easy for the supplied software to display the current pricing information for both the customer and the company. However, resetting prices depends on multiple other factors that must be considered besides simply at the company gross profit margin. Other factors can include elements such as:

- Ensuring gross profit margin for all products exceeds X%. This percentage can vary by customer based on sales volume, etc.
- Cost of providing the orders. The extra cost of running a truck to a distant customer's store. Or the number of stores which deliveries are made to.
- Past performance of promotions. How much did sales increase during a sales promotion and is it enough to make up a lower profit margin? Of course, this also feeds into how many items the customer should order.
- Promotion and advertising rebates. How much is the company going to pay the customer for advertising the product?
- The sales force is working on a commission, thus it is in their personal interest (which may conflict with the company interest) to maximize the figures on which their commission is based.

Making all this work requires more than just a list of numbers. It also needs more than giving the sales force a spreadsheet and letting them play "what if" games. When they are talking to a client they don't have time to work a spreadsheet. Complicating the issue is that each client is motivated in a different way. Rather than providing one set of "what if" spreadsheet values, the system should be customizable by the sales force and the company with rules that adapt it to individual customers or buyers for a customer.

### *Design characteristics*

The following characteristics distinguish complex situations. This section looks at each and considers how they apply to the sales force using the pricing system.

No single answer

There is no single way to determine how to make a pricing change or how much to make it for. Depending on the customer's current volume and current price margin of the item, the price could vary. Balancing against the need for making a profit is the need to keep the customer happy. It might be best to sell the item at cost with the intent of making sufficient profit on other items.

Open-ended questions

As part of having no single answer, the questions have no clear point at which any pricing question can be considered answered. Price negotiation is a constant balancing act between the retail customer and the sales force.

For instance, the customer wants to participate in a manufacturer promotion which changes the wholesale cost, or maybe does not effect the wholesale cost, but the retail store gets a direct rebate after the promotion.

Multidimensional strategies

The strategies of what information a customer responds when making a buying or price change decision varies depending on how they view the situation. The sales force must have enough information to point out the benefits of a view favorable to the company and still providing benefit to the customer. Which benefits look best to a particular customer vary based on both individual buyer and the type of retail store.

Has a history

Price changes have to be tracked so their effect on sales can be anticipated. The system needs to capture that history to provide both what the company and customer need now to make pricing decisions and to modify the presentation to adjust for actual changes in the near future.

Dynamic information

Besides lacking a single answer, the overall situation is dynamic. A manufacturer may start a national sales campaign or offer dealer rebates based on sales volume. Plus, the response of other stores must be anticipated. Starting a new promotion might result in the competition matching the price, and thus, overall sales will not increase.

| | |
|---|---|
| Nonlinear response | The conditions that trigger what pricing information a person wants are unpredictable. As they gain a better understanding of the situation, they might want a highly detailed information about one area and decide to ignore another. |

## *Designer and writer issues*

Chapter 7 describes a set of steps to consider when designing for a complex situation. This section looks at those steps as they apply to supporting sales staff.

### Scope

Many systems have a scope of "provide product prices" and ignore any issues of how those prices are used or the business rules that the sales staff must apply. By working from an overall scope of providing quality, integrated information, the system can work to integrate the information into the methods a sales person uses to interact with customers.

### User groups

The user groups can be much more varied that appears at first glance. Rather than considering all sales people the same, consider how they differ. Besides years of experience there are factors such as account type (national chains vs. single stores) and type of area served (California vs. Arkansas). There can also be human characteristics such as type of personality that influence how they want to interact with the system and how they interact with the customer (the situation).

### User goals

User goal are a mix of both the sales person individual goals of high sales and the corporate goal of high profits. Because each product has a different profit margin, the sales people need to have a clear picture of what products to push. Other goals would be rapid quote generation.

### Information needs

The user goals drive the information needs. Generating the quotes may depend upon past account performance. For example, a high volume account gets deeper promotional discounts or an account that is routinely late in paying gets minimal discounts. Business rules controlling the discounts should also be enforced and quickly obtainable since a sales person may know remember a rule changed or the exact rules for a current promotion.

### Information relationships

Among others, business rules, account history, previous sales performance of the product in the local area, and upcoming events all form an interconnected web of interactions that influence how to best maximize both company and customer profits. These need to be combined into an integrated view rather than just tabular price lists and performance reports.

### Presentation

Balancing the information needs against the user groups is major difficulty in supporting people doing nonprocedural tasks in their day to day work. The basic task of a making a sales call and generating the follow up responses follow a generally fixed pattern, but the details of each sales call are different. The early analysis needs to understand how those details might vary and ensure the system supports them. A dynamic system can provide the different levels of details and even change the presentation style, but the interface design must allow for rapid and easy shifting between different levels.

The system should not attempt to take over the sales person's job, but just provide work as a high quality assistant by providing the needed information when needed and formatted properly. This is the type of presentation Rockley (2001) defined as stage 4 single sourcing. This means the same information, such as promotion rules, may need to appear in different form at different times. For example, the promotion rules that apply to a particular product should be available from that product's quote information. The link should not just display all the promotion rules, leaving the sales person to figure out which ones apply.

# 6 System

*Devices for complex tasks must of themselves be complex, but they can still be easy to use if the devices are properly designed so that they fit naturally into the task. When this is done, learn the task and you know the device.*
*— Donald Norman.*

This book primarily focuses on how to figure out what goals a person sets and what information that person needs with respect to a complex situation. It further considers the factors which influence how the information is interpreted with respect to the current situation. This chapter considers the issues relevant to converting the design and analysis of a person's goals and information needs into a working system that matches that person's needs. It tries to bridge between the theory of the previous chapters and the real-world issues faced by both the writers and programmers involved in implementing a system.

Within the complex situation model, the system provides users with their main information source. Although the system essentially sits outside the situation, the interaction between the person and the system influences how the situation is perceived and understood. This chapter looks at a web-based information system as it relates to the complex situation, and

- Considers that any information system is a tool to be used as part of understanding a problem or situation, and not as an end-in-itself final product.
- Discusses how the system should interact with the user and how it delivers the information. Back end database operations and code design issues are outside of the scope of this book.
- Examines how the most important element in system design is meeting the communication of information to people and ensuring those communication needs match how the system operates.

In accomplishing these three points, it takes three different views of the system: content development, human-computer interaction, and software/hardware design.

## Computer system is a tool

The Norman (1998) quote opening this chapter is often violated because the device (or the software for the systems most relevant to this book,) fails to fit naturally into the task. Instead, it violates at least one of two important points. (1) it treats the problem as a simple situation when it should be treated as a complex situation. The large volume of information to be sorted and integrated causes scaling problems which impair the user. (2) It uses technology for technologies sake. Features such as animation or visualization, which can be useful for some situations, are force fit into all situations. Part of the complaints about Microsoft Word's features that help you write arise from this type of behavior, such as assuming a list whenever a paragraph starts with "1." or "A." or turning any URL or email address into a hyperlink. For being helpful, it's amazing how many people either turn those features off, or complain and never realize they can be turned off.

This chapter considers how the information systems should work to provide a user with high quality information that support complex situations. The focus, as with the rest of this book, is on communicating information to the reader and not software or programming concerns. The computer system is a tool, not a goal, a distinction Constantine (1999) clearly makes.

> All software systems, from operating systems and languages to data entry and decision support applications, are just tools. End users want from the tools we engineer for them much the same as what we expect from the tools we use. They want systems that are easy to learn and easy to use and that help them do their work. They want software that doesn't slow them down, that doesn't trick or confuse them, that doesn't make it easier to make mistakes or harder to finish the job.

The development of an information system constructs a tool which contains some quantity of information the user needs. Depending on the quality of both the system and the information, one of four cases can result.

Good tool—poor information

> The system is easy to use and it provides good support for finding data. But once the user finds the data, there is little support for integrating it into information or the users are disappointed in the lack of content.

Poor tool—good information

> The information may be complete and directly addresses the user's problems. However, because of the system design, the user gives up without finding it or has to perform extra work to integrate the disjoint information elements.

Poor tool—poor information

Information can't be found easily and what information the user does find rarely is enough to address their needs.

Good tool—good information

The system is easy to use and matches the user expectations. The provided information is integrated, fits cleanly into the user's mental model, and is easy to interpret with respect to the situation.

Only the last of these four combinations provide a highly usable system. Regardless of which combination resulted in the poor system, it contains hard to find information or such incomplete information or incoherent presentation that people stop using it because it fails to address their goals and information needs. In the end, many systems often fail to fulfill user expectations and leaves people struggling. Unfortunately, most people have already encountered poor informational systems and are leery of placing any faith in a new system, assuming it to be as inadequate as its predecessor. Thus, when constructing the tool, we must take the proper view; the system operates as a tool that provides integrated, coherent information, not as a tool that pulls text and numbers from a database and displays them.

Actually, when working with complex situations, the designer must remember that a user does not want to use an information system (or any other system); a user wants information. The information system just happens to provide the best method of obtaining that information. Although I acknowledge that a computer system will be supplying the information, this book considers computing to be about providing and communicating information to the user, not the underlying algorithms or data design. Anything that moves people nearer to the information they need enhances communication. The quality of the information system comes directly from the designer's skill in analyzing the user's goals and information needs, and determining the rhetorical aspects of presenting the information to address those goals and needs, not from software bells and whistles or low numbers of bug reports.

### *Software is not the solution*

While the software is part of a solution, it is not *the* solution. Yes, software drives the adaptive system which supplies the information. Systems for complex situations involve users interacting with software to obtain information. An adaptive hypertext system is a highly complicated software system with multiple databases interacting with an interface engine building a custom presentation. However, the important element that determines overall quality is whether or not the person can obtain the information they require, not the quality of software which supplies it. Poorly designed, but bug free software is not going to result in a high quality system. A software system must be viewed as a tool that helps a person understand a problem or achieve a goal, not as a method to display text and numbers. And support for achieving a goal is what users require from the system—not just a bunch of numbers (Constantine & Lockwood, 1999). The difference between the two may seem subtle, but the underlying goal shifts from how to display a bunch of text and numbers on a screen but how to display that text and numbers so they provide information toward achieving a user's goal. The difference between these two views requires a very different design mindset.

Not too long ago, most of the software code and system integration was built from scratch, now a preponderance of vendors are ready to provide their wares. For a short time, all web pages were coded from scratch, now tools such as Dreamweaver and FrontPage provide a WISISYG interface. Meanwhile, software vendors claim their software product is the long searched for silver bullet that will solve everyone's information problems; merely by purchasing the software, all information problems end (McGovern, 2003b). However, out-of-the-box software solutions lack a situational focus and address simple problems, not complex ones; a difference that often results in frustration when the promised efficiencies do not materialize.

Unfortunately, technology drives the current computer market rather than the need for effective communication. Software design tends to put "how to implement it" above "why do it." For instance, Liddle, Campbell, and Crawford (1999) have worked on methods of extracting information from the unstructured information

### System can inhibit or help users

Black (1990) performed an experiment where he compared the performance of graphic design tasks between using pencil and paper and a computer. Two interesting points were found: more time was spent working with computer than with pencil and paper, and the quality of the final product created on the computer was lower. When working on the computer, the people compromised in how they performed tasks and adjusted the goals so that they were easy to perform on the computer. The computer was inhibiting, rather than helping, the user.

Norman (1993) extends this conclusion well beyond graphic design problems to a generalization about system design in general. Many of the problems found in computer systems arise because the system fails to truly support the user and instead, in forcing the user to adapt to the system, encourages the user to take the easiest system-based route. This mismatch between the achieving a high quality solution and the easiest system-based route both increases errors and decreases productivity (Landauer, 1995). Of course, the Black's results can be can be viewed from the other side. With proper analysis and design, the system can ensure that the major work characteristics are easy to perform and thus 'suck in' the users by cleanly supporting their work. "Any system imposes a model of work. the only choice designers have is whether they will design that work model explicitly to support the user or whether they will allow it to be the accidental result of the technical decision they make." (Beyer & Holtzblatt, 1998, p. 7) contained in business reports, but the emphasis was on the extraction algorithms rather than on defining what information is worth extracting and why. As basic research, their approach is acceptable and needed; rather, it's in the application that it becomes problematical. Marchionini (1995) clearly brings out the distinction between the technology and the need for clear communication.

> Technology has changed the way that we represent and share expressions of thought but has not thus far changed the fundamental forms and rules of language and communication. The basic problem of conceptualizing an information problem and translating information from the external world to solve that problem has not been much affected by technology. (p. 180)

In other words, the method of information presentation has changed, but the human mental processing of that information has not changed. As software interfaces present more information and more options for processing the information, considerations of designing for the cognitive processes of the user get swamped by the technology considerations. Software alone does not and never will equal a system capable of handling complex situations. Software might be a vital piece, but it's the information content contained within the databases and the way that information is presented that makes or breaks the system. Software provides a vehicle for delivering boxes to a user and may even wrap the boxes up with pretty bows, but it does not ensure the boxes contain the proper contents. And high quality systems only result when they have high quality and applicable content (Glick-Smith, 2000).

In the end, the problem of providing quality information content will never be solved by technology. Solving that problem requires a careful analysis the user goals and information needs, and the design of systems which supply the information in a manner which meets those needs. Technology can support that process, but alone it is insufficient to make it work (McGovern, 2001).

## *Well-engineered does not necessarily equal usable*

A fundamental goal and quality metric of developing a well-engineered software system is the production of error-free code. A laudable goal since no system can operate well unless the code is essentially error free. However, production of error-free code in no way ensures the system meets the customers' work requirements or cognitive needs (Neerincx & Griffioen, 1996). I've been involved with many systems (some of which contribute to the horror stories in various sidebars in this book) that were technologically sound implementations, but HCI nightmares. Basically bug-free and containing all the requested functionality, they were also almost impossible to use effectively.

Thus far in this chapter, it may should like I'm very negative toward software engineering, but that is not true. My negative feels are not against software engineering, but against extending it too far into the interface. Software internals and software interfaces are two very different beasts. Although software engineering techniques work well for program internals, they are problematical for interface design (Borenstein, 1991; Diaper, 1997). .Bailey (1989) claims many designers work at the overall problem backwards. Rather than defining system objectives and then working to meet those objectives, they create a basic design of the system and then set the objectives and performance specifications. He considers two problems with this approach. First, the objectives no longer focus on the user but, instead, fit the preconceived technology, and second, the approach discourages different design approaches. Both problems harm the usability of the resulting system and contribute to the estimated 50%-plus failure rate on major software projects.

## Content development

This section of the chapter looks at practical issues of creating the content. Since I'm not addressing a particular methodology or development environment, the best this section can do is provide big brush stokes that point toward a solution. Each implementation will have to define the specific details.

Sprague (1995) claims the major value of online information lies not in providing management of database records but in communicating the concepts and ideas within that information. Delivering that value means the goal should be efficient communication with people who receive that information and not on efficient storage and data manipulation. Granted, efficient storage and data manipulation is very important to ensuring the system can deliver the information, but they are not an end in themselves. Instead, they contribute to a system for complex information that anticipates the users' needs and derives the users' high level goals based on user actions and responses (Nardi, Miller, & Wright, 1998). What the reader sees is not a document that has been carefully groomed by a writer and editor, but rather a document which was compiled from a database just before the information was presented to the reader (Albers, 2000).

### People versus "performance and technical considerations"

Well engineered systems designs often ignored that the system design contains a human and must reflect concerns with human performance as well as machine performance. Many systems have been designed to maximize the machine performance with little regard for human performance, or with the assumption that the human component would adapt. A design attitude which leads to the overused lines: "that's a training issue" or "explain it in the documentation." I worked for one company that had a rather well engineered system that was also user hostile as far as assimilating the information it contained. Considering that this was an electronic medical patient record system intended for use by nurses and physicians, I was rather concerned. Those health care professionals might be making major decisions about MY health. The programming manager dismissed essentially all suggestions for HCI design changes because of "*Performance and technical considerations*." (Yes, you could hear the italics intoned in her voice.)

### No low hanging fruit

Design which supports understanding complex situations magnifies the need for methods which can handle both the user's initial collection of information and its continuing refinement. Rein, McCue, and Slein (1997) emphasize that the "reader is first a seeker of information" (p. 85), but after the information is found, then it must be refined and manipulated to be understood within the situation context. Providing a system that helps users find and transform data to meet their information needs requires deeper understanding of the user and the real-world questions the users want to solve (Albers, 1997).

A common design problem is addressing the easy issues first (low hanging fruit) and ignoring more complex issues or setting them aside for a later release. However, a design that has implemented an approach that emphasizes low hanging fruit aspects of content does not lend itself to more complex issues. At a later point in development, faced with a deciding on whether to do complete redesign, too many companies opt to keep the low hanging fruit approach. The design team gets that wonderfully conflicting set of messages that people are complaining the system isn't useful, so fix it, but don't change anything because the users know the current design and changes would confuse them.

In this context, low hanging fruit are the easy to design or implement factors which, unfortunately, often do not scale up. A couple of common examples are FAQ or other question and answer lists and the use of multiple tabs to sort information (figure 2.4 on page 64 provides one example). Easy to implement, these work for addressing simple situations, but they fail to scale for complex situations. They do not scale because they assume essentially a flat information hierarchy and that each element is independent of all the other information. As I've mentioned enough that it has almost become a mantra, a user never just needs more information available (i.e., more questions and answers, or clicking between tabs), but an integrated presentation that relates the information to the current situational concerns. No hierarchy of questions sets of tabs will address that need; the lack of relationship between questions forms a fundamental barrier that requires a completely different approach.

The real design problem is not addressing the low hanging fruit part of the content. It's clearly understanding the requirements for covering the entire communication aspects users need to work within a situation. The initial design plan needs to lay out how the final, full-implemented system works to meet user's complex goals and information needs. Then features implemented early in during development must be balanced against that plan to ensure it meets long-term goals, rather than simply meeting short-term, initial design goals.

### XML for content markup

I am working from the assumption that the textual data will be developed in XML and stored in an XML database. I make this assumption because XML technology seems our current best bet for designing systems that handle unstructured (non-relational) information and for providing high quality output in multiple formats.

Articles on structured documents via XML and SGML, with individual components stored in document databases, paint picture of multiple writers at multiple locations contributing information which feed into a single sourcing system to dynamically generated into a document.

On the other hand, XML is not a hard and fast requirement. Rather, the hard and fast requirement is to communicate information to the user. The designer needs to work with the rest of the development team to choose the proper method of storing and delivering the information. Focusing early design efforts building a visual diagram of the goal and information relationships and the information model (Hackos, 2002) gives a structure against which to map the information storage requirements. Ensuring the information fits within the current reader context and is relevant to that context must be the designer's main concern.

From both the technology people and the communication people, XML, and content management systems are being touted as the current best answer. The recent reports of design methodologies all assume use of some combination of these XML and content management, but they are often based on creating static documents with multiple outputs and not for dynamic information (Hackos, 2002; Rockley, 2002). Whereas XML would be my current choice to develop a system for complex information, I also must acknowledge that I expect a host of unforeseen problems to crop up around controlling and scaling up content design and development that would have to be solved (Albers, 2000). Actually, I expect these problems with any system using multiple writers to generate dynamic information. In the end, as I have previously mentioned, we must not lose sight that although these technologies might allow for constructing a good solution, they are tool sets, not a solution.

### *Creating text elements*

The text must be created in small pieces. Each piece, a text element, may run from 50 to 300 words and include the information required for any level of knowledge, detail, or cognitive ability (see "Multiple dimensions of a user's understanding" on page 51). When the text elements are assembled into the final document, the actual content will be adjusted and the final word count could range from nothing to most of the words. Text development involves both writing the text and inserting the proper markup to enable the text modification.

Most current single sourcing develops text for multiple outputs or multiple products. Thus, the conditions tend to be print/online or product 1/product 2. A simple list of different output options can be produced and the final result for each product edited. However, in a dynamic system the conditionals change; they change the amount of background information or detail level of the presentation to match the user's needs. With multiple variables, the large number of different potential results and time constraints prevent editing and reviewing each individual result. Plus, as a user interacts with a system, the information presentation can change with time.

Developing text in this manner is difficult. One large software company found their writers have trouble working at less than a chapter level of a software manual (Jones, 1999). Unless this problem is symptomatic of new process and not a long-term writer issue, writing for dynamic text could prove very difficult. For dynamic complex information, the writing needs to be done at a much lower level than a chapter. Currently, we neither have the design methodology for defining what text element contents, nor do writers have experience in writing disconnected small pieces. Additional issues include figure out how to handle transitions to seamlessly combine these text elements into coherent documents. A document that sounds like a bunch of individually written paragraphs stuck together will not effectively communicate. Defining the content and writing text elements requires a focused research program; one which this book only scratches the surface of addressing.

### *Complex systems are not rule based*

Almost by definition, a rule-based system means a one-size-fits-all system, or at least, following a path provides a fixed terminal point. For simple systems this works wonderfully, and for complicated system this works well, with the main problem being overall size renders it unwieldy. When all the paths through a situation can be defined, then a rule-based system is an optimal design choice. If rules could be defined, Clippy from Microsoft Word and other assistants like it would be a highly prized addition to a software program, not the butt of jokes and turned off on essentially all computers. Likewise for the other auto-formatting features of Word, such as auto-numbering when a paragraph starts with a number. When the writing fits the rule they were designed for, they work great. When the writing does not fit the rule, they interfere with the writing process and can cause great problems for people that don't understand how to undo the "helpfulness." Writing is too dynamic and has too many variations to correspond to a rule-based system.

In chapter 3, I briefly discussed wizards as rule-based methods of walking people through simple situations. They work for installing software, but fail for car buying or providing medical information. A common statement about wizards is that they let a user get close and then it can be tweaked. (I've repeatedly heard this about the Excel chart wizard.) However, for complex information, the needed tweaking can be hidden; a user needs to know what to tweak and why to bother tweaking. Whereas for a chart, the desire to change a font is obvious, the need to adjust possibly misleading graph scales are not. If the user didn't realize that the fonts or scales could be changed, they would have no basis to ask the system to provide that information. Likewise, the need to request more information (dynamically expanding the detail level) or knowing which information to look at in more detail is often not obvious to the reader who does not already understand the situation. Although a business analyst might understand the overall corporate situation enough to know which numbers to examine more closely, a person learning about a topic lacks that deep knowledge that allows for asking follow-up questions. Addressing a reader's knowledge level means knowing how to put some of the follow-up information into the initial presentation and what can wait until later in a manner that fits each reader.

Rules fail with complex systems because the rule-based design fails to truly address meeting the user's goals. The fundamental problem is that people normally don't know what information they truly need. Also, the open-end nature of questions prevents predefining what is needed for a specific situation. A rule-based system may get the person into the ballpark, but it fails to provide final coherent resolution to the goal and information needs. Getting a person close can be a problem since they may accept close enough as an answer (or not know how to go about getting closer) when, in reality, the answer is still inadequate.

Rules are algorithms, but algorithms don't work within the world of dynamic relationships where the weights vary by situation and by individual. Basden and Hibberd (1996) believe that both collecting the information and using it are knowledge-generation activities and that often the questions which arise are new questions that users could not have considered in advance. Systems that allow for dynamic manipulation of information support the user in answering new or revised questions. On the other hand, rule-based systems require all questions to have be considered up front as part of content creation. The emergent properties of complex systems makes it impossible to predict what information is needed and how the information is needed. When the information needs emerge as part of how user's build and perceive information relationships then rather than through fixed rules, the system manipulation involves modifying the changing relationship parameters and the associated information. A major part of clearly understanding the user goals and information needs is collecting and interpreting the audience analysis to understand how those relationships might fit together (Cilliers, 1998).

Any complex information system that is suppose to provide information cannot be rule-based. Missing from rule-based systems are several factors which are critical to providing coherent information in complex situations.

| | |
|---|---|
| Ways to combine related information | Almost by definition, a rule-based system has an underlying hierarchy of information and the system walks the user down a particular path. Missing is the ability to combine some of the information on two or more paths. Think of the number of times a computer system has asked you some binary question (yes or no) and the real answer is maybe. As a user, you need the applicable maybe information to be combined with other maybe information. |
| Ways to expand or contract information | Information that is already known should be condensed to provide a reminder, but not overwhelm the new information. Alternatively, new information may need more details. It may need to expand to provide underlying details expounding on the information. The entire presentation may need to be reconstructed to present those underlying details as part of the integrated presentation. Later, this information can be presented with a shorter reminder. |

## *Rules assume a 1:1 mapping*

The technology-driven camp focuses on software as a solution and tends to ignore the rhetorical and human aspects (probably because many computer science education programs do not provide adequate training in the rhetorical and psychological aspects of human-computer interaction). Yet, many designs that should address complex information needs fail because they assume that technology can adequately address the problem; a view Norman (1998) has dedicated a book (*The Invisible Computer*) to criticizing. As he points out:

> The proposals are always technical solutions, whereas the problems reside within the person; cognitive tool are needed to aid in the programming task, social and organizational tools are needed for the group problems. And these human problems are harder to solve than mere technical ones. (p. 95)

The subtle details of uncovering when and why a user wants information becomes a design stumbling block. Part of the problem stems from a focus on the easy-to-see aspects of the design, as Vincente (1999) points out:

> Much more attention was paid to the syntactic and lexical features of the interface (e.g., Should windows be titled or overlapping? How broad and deep should a menu be?) than to the semantics of the application domain. This emphasis could also be seen in the importance placed on usability—design computer-based system that are easy for workers to use—and the comparative lack of attention to usefulness—designing the functionality that is required to do a particular job effectively. (p. 91)

Notice how the important aspects of the preceding paragraphs were people, not technology. A technology view revolves around a belief that, with the right software, most content management and delivery problems could be solved in a cheap and efficient manner. A "people view," taken by this book, revolves around a belief that content is, at heart, about people writing, editing and publishing content for other people to read and use, with technology playing a supporting role.

As I discussed in chapter 2, a common problem is to try to simplify a complex problem and define it as a simple problem. In the process, issues such as mapping the audience analysis to information content can appear to be a simple problem: each requirement identified in the analysis can be mapped to one item of content. By bringing in a basic application of single sourcing, information reuse can even simplify the content management problem even more. The simple mapping is the low-hanging fruit syndrome. The mapping is easy to implement and verify. On the other hand, the mapping does not scale up.

However, this approach has two major issues that complicate the implementation. First, it ignores the dynamic nature of people information needs. The important aspect of understanding a complex situation comes from understanding information relationships, not the information itself. Second, with a complex situation and the large amount of dynamic information being processes, issues with scaling. The highly interwoven web of information relationships is not a linear scaling from a simple system. The information presentation required to meet user needs can cause the system to take on some non-linear aspects with only small changes in initial conditions requiring very different final solutions.

In chapter 2, "Multiple dimensions of a user's understanding," explained how knowledge and detail levels are both important components of a user understanding a situation. The problem of mapping an audience analysis to either one (let's say knowledge) is that neither is a single enitity. A person's knowledge about the various subtopics within a topic varies. Thus rather than having a single value K, there are many $k_1, k_2, \ldots, k_n$ that need to be addressed and the values for each $k_a$ for each person will vary. To greatly complicate the problem, values for each $k_a$ can change when interacting with the system and gaining an understanding of the situation. There are too many permutations to predefine the information anyone person needs before hand, but a adaptive system that picks matching elements can dynamically build an appropriate document.

## *Metadata*

Metadata provides the connection between the information in the database and the information retrieval and adaptive software. The adaptive system can process the text based on the metadata contained in the text tags (assuming XML) to produce the final output.

When used with text, metadata is too often limited to categorizing the text with values such as revision date, author, and department. Important information for maintaining high quality and current content, but not helpful for dynamically creating documents.

For an XML implementation, the metadata should exist as attributes on the tags. For example, in the three dimensional model in "Multiple dimensions of a user's understanding" on page 51 tags might have attributes that match each axis. For example, <para knowledge="3" detail="4-8">. Other metadata attributes might take into account the various characteristics of a complex situation such as history. Of course, these need to be balanced against the profile collection system being used by the adaptive system.

The tagging of content with its metadata must be done by the writer while the text is being written. There is too much contextual knowledge in the writer's head that will have long since been lost if a two pass system of write text and then tag. Of course, in many cases, it'll be subject matter experts writing the first draft. Teaching them the nuances of writing text elements and conforming to a metadata system will often be counter-productive. However, what the subject matter expert writes is just a first draft, a writer still needs to break the text down into the text elements and properly tag it. Writing and thinking in terms of text elements is not easy; add the requirement for multiple audiences and the writing task becomes much more complex (McGovern, 2003a).

## A stereotypical bureaucracy

A good example to contrast to a rule-based system to an adaptive system is a stereotypical bureaucracy. Consider the terms that most people use to describe a bureaucracy: inflexible, unbending, procedure-oriented, fixed. All of which stem from an underlying attempt to prevent any evolution or adaptation. Every process is carefully defined and each step in a process must be followed or the element being processed is rejected or returned to the failing step to be fixed. (Think of all the movies scenes where the stone-faced middle aged lady deadpans "I'm sorry, you didn't include form 3942 stroke G3. You must resubmit your complete application. Current processing time is about 6 months. Next in line, please.) Unfortunately, the highly regimented bureaucratic view fits very nicely into highly structured system methodology, which can lead to designs that unconsciously attempt to mimic it. (A methodology which works fine for simple situations; the problem occurs with complex situations.) Rules must be defined for each piece of information, which allow for control and structure. The main problem is that the rules become almost impossible to change although the situation which lead to their development may no longer be relevant. In the worse case, rules freeze some of very bureaucratic overhead which the new system was to eliminate.

The brittleness of the system can be seen in two different common bureaucratic actions. The first is the constant shuffling of something which does not conform exactly to the process rules. Rather than acting on the element which fails to fit the rules, it is sent to someone else; in a computerized bureaucracy, it must go to the exception handling group. In a computer system, the shuffling from office to office can't actually occur, but the user can be left cycling through various places within the software system trying to find information which does not exist. The second is workarounds. People who frequently work within the system quickly develop methods and techniques to bypass the established rules. ("I'll just take this to Jenny and we'll have it approved tomorrow.")

Missing from the carefully structured information (and the bureaucracy) is the most important part, the dynamic nature of relationships between information. Those relationships do not lend themselves to a rule-based description and, thus, fail to make it into many system designs. It's not that an attempt was not made to incorporate the relationships, but that the attempt focused on defining each identified relationship as a fixed rule. (The need for an exception handling group points out the failure of capturing the rules.) The interesting point here is the almost universal agreement on the frustration felt by a person forced to work through a bureaucracy to obtain something (and creates the humor of the situation with the stone-faced bureaucrat in the earlier example). The frustration of a person finding something within an overly ridge information system also exists, but often gets rationalized away as a training issue. People are extremely adaptable and will learn to use, if forced, even the most user-hostile systems. But in most cases, user frustration is not a training issue, but a design issue.

An approach which uses metadata within the XML tags should provide more flexibility than an approach based on conditionals (IF statements). The main problem with conditionals is they are all embedded within the text itself and can become both a maintenance nightmare and harder to reuse. Reuse problems would crop up when the variables tested within the conditional are not defined. Although it may be easier to initially write text using conditionals, the text element will probably be longer and more limited in its possible reuse.

## Human-computer interaction

A primary assumption about adaptive systems is that they will be producing web-based hypertext, and this section works from that assumption. However, nothing actually forces this limitation. The system could produce a customized document that is printed as part of the production process. For example, in a healthcare setting, printed documents could be given to the patient as part of the patient education and those documents can contain information directly applicable to the patient. Such as addressing medical concerns in a way the incorporates both the current medical problem and a pre-existing heart problem.

### *Information architecture*

The Web is inherently a navigational space, readers like a consistent navigational architecture. The existing principles of information architecture still all apply, such as those laid out by Rosenfeld and Morville (1998).

### *Delivering information to users*

Designing the text to address a complex situation requires much more than defining a clear architecture. In complex situations, the system operates as the user's main information source, but only as the main source, not the sole source. The major distinguishing factor of whether the information is delivered effectively or is just a dump of data depends on how well the system providing the information addresses the following questions. Whereas these questions drive content issues addresses in the other chapters of this book, this chapter looks at ways of taking the answers to these questions and deciding how the delivery vehicle works.

- How does the person expect to access the information?
- What information do they expect to find?
- What do they expect to do with the information?
- What relationships exist that connect the information to the real-world situation.
- How does the information system achieve (or fail to achieve) its purpose within that relationship of information to real-world situation?

According to Parker, Roast, and Siddiqi (1997), part of the problem comes from the interface design being subordinate to and defined later than program core functionality. Bodker (1991) states that "both practically and theoretically, that to write computer programs and to describe human work, are two different things, and that formal descriptions are perhaps not the solution in the latter case" (p. 6). Recognition of the difference and attempts at handling the problem has lead to creation of user-centered design methodologies. Hakiel (1997) points out that the domains of software engineering and usability engineering are very different in how they consider usability and context. Karat (1995) found developers consider a problem solved if the information is found, while people don't consider a problem solved until they can make use of the information. Bailey (1993) found significant usability differences between designs produced by human factors specialists and programmers. In informational systems, these problems can create major problems since context dominates the system use, meaning information design considerations must take priority over programming decisions. "Products do not just have to work reliably, they have to do the right things in the right contexts" (Rouse & Valusek, 1993, p. 524). Neglect of the end user issues produce systems which, while they may be well-engineered and bug free, do not work in the expected manner and, thus, do not fit coherently into the complex situation faced by the user (Cohill, 1991).

### People adjust to match the system

The design and presentation of the system drives how a person learns about and understands a situation. As the main source of information, it normally exerts a strong influence over how the situation evolves since it drives how a person perceives the situation. Thus, it is important to consider how the system development how it operates as a communication tool. A substantial problem with many systems is that the underlying tools used to develop the system drove the design decisions for the human-computer interaction of the system. At best, this is a questionable method and, at worst, it results in usability disasters with an efficiently implemented (from the development tool viewpoint) but impossible to use tool.

People can learn to use any system. Whether they will depends on their need levels, such as having to use a bad system for work-related tasks. More importantly, for understanding complex situations, a person's mental model develops along lines that match the presentation. If the person has a better or more realistic mental model which conflicts with the system design, then they either create work-arounds or feel trapped by system limitations they do not provide information they need.

The problem with having people adapt to a system, rather than the other way around comes from various issues of how people handle information. They find it difficult to effectively search for and integrate multiple sources of information (Lansdale & Ormerod, 1994), something a poorly designed system requires at essentially every interaction. Also, how information gets presented can cause people to make opposite choices and believe them to be best (Tversky & Kahneman, 1981; Johnson, Payne, & Bettman, 1988). So poor presentations can lead to people completely misunderstanding the situation and feeling confident in their understanding.

## *Filtered and processed information*

The fundamental reason people access an informational system for complex situations is to collect sufficient information to understand it and, if required, to track the development of the situation over time. Murray (1995) considers the problem for the viewpoint of helping business executives:

> Executive do not need help with decision making; they need help with gathering information. And, this could happen if executive used Computer-mediated Communication to gather information from a broad base of employees, and if they used appropriate information retrieval systems. But,... these information systems may not provide the filtered and processed information the executive really needs. (p. 49)

"Filtered and processed" is the key phrase in the above quotation. Supplying the filtered and processed information requires that the writers and designers who supply the content have a clear grasp of the executive's goals and information needs. For a user to know when sufficient information has been collected requires gaining a understanding of the problem's entire situation. Preventing information overload requires the proper amount of filtering and processing: long pick lists of titles or lists of initial document paragraphs do not qualify.

In today's world, the information will almost invariably come from a database that exists as part of large computer system. A major failing of older systems was that they never had information designers involved. Instead, the programmers took the internal structure of the software or database and reflected it in the outputs: the user interface and printed reports. If the software is talking to other software, then an interface which reflects the internal structure is okay; that is a highly efficient way to program. But when a person gets involved and has to work with the interface, the rules change. The design must focus on effectively communicating with the person interacting with the system. While an important component, data movement from database to screen is secondary to communication between the screen and the user. Make the system do the work, not the person.

With the rise of the Web, there was a partial shift in attitudes toward making the information accessible. A major reason was because putting information into the interface became the major reason for it's existence. Brochure-ware web sites assume a simple situation model although the products or services they describe might be better answered with a complex model. Luckily, web design as brochure-ware has been rejected as quality design. However, even with acknowledging the need for people to get to large amounts of information, the design still more often supports an efficient pipeline from database to screen information system than it addresses the user's contextual situation. Single sourced databases are used to fill in simple templates that are but a step removed from brochure-ware. Or a strict hierarchy of information is defined with no overlap or minimal reuse of information between pages. In a complex situation, user's might need to see same information in conjunction with many different pieces of information. Single sourcing provides a good tool to accomplish that; the design analysis must figure out when to provide it.

The technology used to communicate the message should not be chosen until the goals and information needs of the audience are defined. Good information designers do not start with a six screen web-based design and then figure out what content to fit into it and who the audience will be. Rather, they start with understanding the goals and information needs of the audience and what information is available, then decide if paper, web-based, or a loudspeaker is the best method of communicating that information. As I wrote this book, sites using Flash and Shockwave were appearing all over the web. Unfortunately, these sites often forget that they have a message to deliver and are little more than a multimedia production of special effects that leaves the reader dazed and fails to deliver a coherent message. In many circumstances, animation can help deliver the message, but in too many situations on the web, the animation was obviously done for the sake of having animation. Dazed readers, suffering from information overload, should never be the result of interacting with a design. Using fancy technology can be highly seductive and fun for the designer, but not necessarily the best for the reader. Stay focused on the reader and avoid building a technologically cool Rube Goldberg machine to communicate the information.

### *Users never jump through hoops*

In any system development, many issues arise that require design decisions that often boil down to "who jumps through the hoop?" Or, should the implementation be the easiest/quickest for the developer or the best for the user? Unfortunately, as the usability literature constantly shows, the decision is often made in favor of the developer. I once worked on a project where the development lead refused to consider a better search method on a customer service application since the user could type Boolean expressions into any number of fields on the screen, click enter, and get all matching records. The users, in general, were older women with only a high school degree; they had no idea what Boolean logic was, much less how to combine expressions across multiple fields.

With a system for complex information, the developer might want to try to simplify or complicate the design. The simplification comes when they try to ensure a one to one mapping between program requirements and internal procedures. This makes it easy to test during development and to ensure all requirements are met. However, lacking a clear knowledge of the problem domain, the real information presentation a user requires does not get tested. On the other hand, the design can be complicated by trying to give too much control to the user. True, I've made many calls for the necessity of customizing, integrating, and manipulating the information, but this is not a user free-for-all. Instead, the design analysis needs to define the potential paths and ensure the user can easily move down them. They do not need access to any and all possible paths they might think up. The end result of a such a system would be a data dump with text manipulating tools; the exact opposite of the integrated information required in a complex situation.

### *No expand or collapse text*

Many current systems allow paragraphs to be expanded or hidden as a method of providing the relevant information. For procedural information, this method works well since experts don't need the supporting information. But for complex situations, rather than simply hiding or showing text, the level and amount of integration should change. The real needed content in the paragraph may need to shift from one sentence to several sentences or the language used might change from technical to non-technical explanation.

Integrated information does not arise from expanding or collapsing paragraphs. As a first level of adaptive design, expanding or collapsing text at the paragraph level works, but it fails to provide for user's varying needs and fails to scale to more complex information. It can be compared to web pages that declare themselves as interactive media because they contain hyperlinks, especially early web pages with their sole navigation being <prev> and <next> buttons at the bottom of the page. Basing an adaptive design on expanding or collapsing text is easy to write since there are usually few conditions which determine the expansion. Instead, a detailed description is written and paragraphs are collapsed as a means of providing the expert with easy access.

A built in assumption is that information relevance can be defined at the paragraph level. But integrated information needs to work at a lower level. Some people need more information within the paragraph than others; for example, term definitions or background. Based on individual needs, information may need to be added, removed, or changed within the paragraph. Most of the adaptive implementations I've seen have be implemented for procedural text in simple situations with the assumption that the expert doesn't need any detailed text. True, but that assumption isn't necessarily true for complex situations. There is not a fixed amount of information anyone needs and different histories can change what a person needs.

A decision to include information within a paragraph needs to be made based on the topic and the individual reader's needs. With a high quality dynamic system, rather than adding or hiding fixed text, additional text manipulation customizes the paragraph text for the individual reader.

## Software and hardware implementation issues

This section describes the basic elements that comprise the current best bet at putting together a system that can address complex situations (Figure 6.1). This section takes a very high level approach since the implementation and design of these elements are not a major concern of the person communicating information for complex situations. Rather, they are the concern of other people on the development team who are specialists in those particular areas. For the purposes of this book, the system can be considered a black box that takes information from the content providers and the situation and deliveries it to the user. For instance, detailed descriptions of information movement within the database and data manipulation within the system are outside of the scope of this book. On the other hand, each element has its own research area and a vast amount of literature. Together, they are all important areas which must be addressed in implementing an enterprise level system.

## *Adaptive hypertext*

The use of adaptive interfaces continues to grow in importance with changes occurring at such a fast rate that any details will be far outdated before this book is printed. Thus, it will only provide an brief overview of the concept and consider how it can be applied to design of information systems for complex situations. Implementation issues will not be addressed.

An overwhelming percentage of major corporate websites are using basic adaptive interface techniques to dynamically generate content, a movement that is sure to continue as more enabling tools for single sourcing (mostly currently revolve around XML) come onto the market. The move to dynamic generation comes from both the need to provide increasing amounts of information and the user's desire for personalized information. Unfortunately, the adaptive techniques currently in use are extremely limited, often little more than populating templates with static text and some level of "display paragraph/hide paragraph" conditional execution. Granted, the static text can be changed rapidly to simplify website maintenance, but this is only scratching the surface. In the longer term, adaptive hypertext researchers want to use both the user's past queries and information in the database to define how the document gets generated; a research agenda that focuses on technology and system internal design.

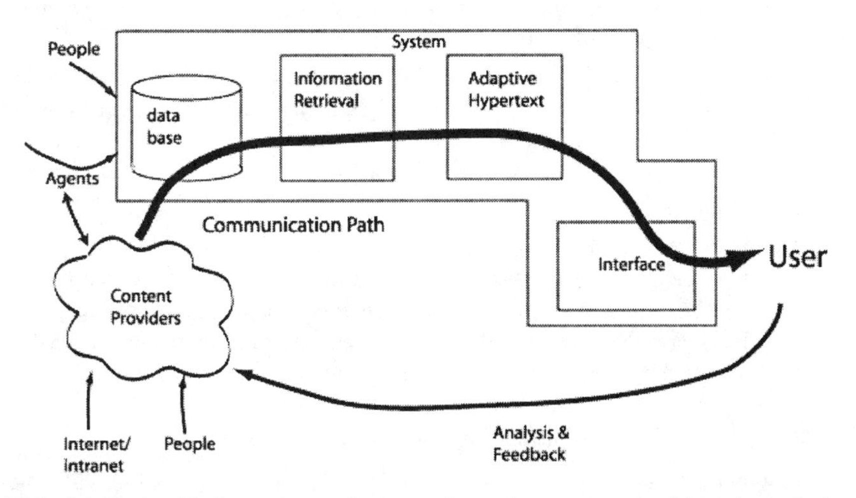


FIG. 6.1. Path of information as it moves from the content providers through the system to the user.

Essentially, adaptive systems "constitute a deep structure to the interface and the aiding output is expressed in the surface structure" (Eggleston, 1997, p. 95). An ideal adaptive system would progress in complexity depending on the level and type of knowledge the user has. Building on research into how people develop information search strategies (Archer, Head, & Yuan, 1996) and react to explanations (Johnson & Johnson, 1993), the adaptive system displays different information elements in a dynamic manner which reflects the user's past interaction with the system. The multi-dimensional analysis I describe in chapter 2, "Multiple dimensions of a user's understanding" on page 51 and in previous work (Albers, 2003b). Adapting the information requires clear understanding of both what information constitutes the dimensions of knowledge and detail level and how to manipulate that information to the situational context.

Computer science researchers are actively pursing research programs in adaptive hypertext and information retrieval to facilitate the rapid and efficient creation of the dynamic documents (Kaplan, Fenwick, & Chen, 1993; Lee, Leong, & Si, 1998; Ponte & Croft, 1998). Most of the adaptive systems try to dynamically construct a model of the user. Examples include:

- Building models of the individual user (Brusilovsky, 1996).
- Creating custom index structures based on system use (Mathe & Chen, 1996).
- Understanding the document structure and adapting it based on the user model it dynamically builds (Boyle & Encarnacion, 1994).
- Generating documents dynamically based on user input (Gruber, Vemuri, & Rice, 1997).
- Monitoring and reducing cognitive load via adaptive techniques (Ray, Hess, & Goldberg, 1997).
- Dynamically expanding or adapting user's information requests (Hancock-Beaulieu & Walker, 1992).

But, for effective communication, the problem needs to be viewed through a different lens. Whereas computer science is rightly concern with algorithms for rapid information retrieval and information selection methods based on user actions, it ignores the fundamental concepts that underlie effective communication of information.

> Users may not be as goal-oriented and rational, as some of the adaptive features of proposed systems require. People will act based on the situation they are in right now, so their goals and plans keep changing in response to how the situation changes and develops. If the adaptive system assumes a too rigid and static model of the user's plans and goals, it will not be able to capture the "continuous improvisation" that people are involved in (Hook, 2000, p. 421).

Adaptive techniques are important because they support the communication process to allow for "increasing the scale of information that users can handle at one time" (Robertson, Card, & Mackinlay, 1993, p. 59). Adaptive systems should leave the interface predictable and not force users to interpret annotations, and should not change the structure of the information space (Hook & Svensson, 1999). An issue which must be studied closely is the mental dissonance caused by unexpected changes to the interface presentation, such as changes in dynamic menus. With an adaptive system the text can change between visits, which can impair peoples ability to refind information. What type of design features will minimize the issue and how to maximize both refinding information and meeting changing user information needs requires more research.

One important issue which often gets overlooked in system design and implementation is that although adaptive hypertext techniques are imperative for highly complex systems with large information spaces, the adaptive hypertext theory does not provide any guidance into what information must be created in the first place. Adaptive interface research often focuses too closely on ensuring it retrieves only the 30 relevant records; what we also need is research into how to write those 30 records. Too many software failures (from general system development, not just adaptive systems) arise because the technology gets divorced from the user goals and information needs and meeting the technology needs drives presentation with actual content being relegated to an afterthought. As a technical communicator, obviously I strongly oppose this view, but do acknowledge that both the content requirements and technology requires must be balanced in order to achieve an effective implementation..

An adaptive interface can assist with the problems of putting together chunks of information. However, unless carefully planned and developed with a clear view of the intended audience, the result can degenerate into an incoherent hodgepodge of semi-related dynamically generated data (Weiss, 1993). Research into this area is only beginning. For example, Benaroch (1996) has looked at how to incorporate explicit knowledge into the design of knowledge-based systems, the very systems that best address complex situations. However, his work focused on the technical aspects of the system. The basic communication aspects remain an open, and minimally researched, question.

### *User profiling*

A basic premise of this book is that a system can provide dynamic customized information to a user. Thus, knowing what information to present is obviously a major concern. Defining and using user profiles is a core element of adaptive hypertext research. Like the other issues of adaptive hypertext, I've ignored the issues of profiling as the computer science researchers are still working on the problem and the level of understanding of the problem is changing rapidly.

The goal of user profiling is to know enough about the user, without the user consciously providing the information, to present the proper information. Based on the user ID and a record of past interactions, the system dynamically determines and maintains a record of the user's knowledge level, what has been previously read, and other pertinent data. That record is then used to influence how the information is presented.

## Database

Most information users view all the information they receive as coming from a single database that is populated by a combination of work by the content providers and software agents pulling information from either the Internet or an intranet. In reality, the data spans many database servers; the textual data and transactional data will almost always reside in different places. However, for the purposes of the system from the viewpoint of this book, the actual storage mechanisms are not a concern. Database administers do a wonderful job of making it all appear to be a single database.

## Software agents

Collecting and refining the content requires a system that can acquire huge amounts of information objects; reuse, render and publish them in different formats for multiple channels, and control their revision and publication. Plus, such a system must also be scalable to accommodate growth. The two current methods best designed to accomplish this are software agents and web services.

Software agents have the capability to automatically search and mine the Internet and retrieve information or to keep existing information updated. Unfortunately, although this is a very promising research area, it's not quite ready for prime time. The work of Blandford and Barnard (1997) on interactions between various agents (including humans) within the system makes a good starting point for starting the analysis of complex situations. Benyon (1993) claimed that adaptive systems should build intelligence into the system and that it is required for very large databases as dynamic pruning of the search space becomes imperative (Kumar, Plaisant, & Shneiderman, 1997).

In most articles on agents, they are portrayed as completely replacing the content provider. They search the Internet or watch certain sites and informing the user of new or updated information on a certain subject. Of course, missing from these articles is consideration of the basic question of where the information comes from in the first place. Who creates the mined information? How does the agent know what information to return and what to ignore? All issues that require the content provider to program the agent according to the user's goals and information needs. Part of the future content providers job will be to control the agents collecting information rather than writing all the text.

In supporting a system that functions according to the ideas in this book rather than directly informing the user, agents help find data that populate the database. The amount information for any non-trivial information need exceeds the capability of a reasonable number of people to generate and maintain it in a reasonable timeframe. By having agents find and perform an initial data sort or verify existing information is current, they allow the content providers to focus on ensuring the data gets properly communicated to the user.

### *Web services*

Along with agents, web services will provide information both to the content providers and directly into the database. Web services are self-contained, modular pieces of programming logic that perform specific functions. From the perspective of the content provider, they differ little from an agent (although from the system engineering and programming view they are quite different) in that they are set up to automatically supply information that fits a set of parameters. A major distinguishing factor is that agents actively search for information and web services provide it on request. So, you must know the web service exists. In other words, for an agricultural site, when you make a request to it, the web service can supply you with a collection of data on corn production from 1900 to the present.

To put this in perspective for a future reader of this book, web services are a current hot topic in the computer trade press as an up and coming technology. Whether the future bears out the current hype and promise remains to be seen.

### *Information retrieval*

Information retrieval takes the input collected by the interface, feeds it into the databases, and collects the data which matches. Unfortunately, the research seems more concerned with problems of information representation within the system and concentrate on back end methods of performing faster searches with better retrieval, rather than focusing on how the information can help users or how to present the information (Blandford & Barnard, 1995; Dattolo & Loia, 1996).

On the other hand, the information retrieval operates inside the system well beyond the view of the user. By occupying that position, the relevance to interface design and interaction declines and the computer science focus on efficiency is the proper focus. The problem comes in systems that provide the user little more than a pile of retrieved data, with search engines providing a good example. Of course, search engines have a very different purpose than a system designed for dynamic content generation. But they do show what happens if the system just provides a list of documents that potentially address a user's goals.

## Conclusion

For complex situations, an information system often is a primary, but not excusive, means of obtaining information. As such it must deliver content which fits the person needs and needs, and which allows integration of that information with other outside sources.

To fit that requirement, an information system is a tool used as part of understanding a problem or situation, and not as the final product. The user requires a robust system that can handle both changing conditions and factors not considered in the original analysis (Cilliers, 1998). Supplying that information in a form fitting both of Cilliers's conditions is the difficult part of addressing complex situations, requiring:

- A deep understanding of the potential situations the system will address.
- A deep understanding of the various user groups goals and information needs.
- A deep understanding of how the people relate the information system and the situation, and how they want the information delivered. Getting a grip on this point forms the basis of the major part of chapter 5.

With those three deep understandings in hand (not a trivial task), then the designer and system engineer are ready to work on the system interaction model and the various back-end models required to support it. As Vicente (1999b) has pointed out:

> Previous experience has shown that the value of any particular design is typically constrained by the thoughtfulness and thoroughness of the analysis that preceded it. If interfaces are to support users in their work, then the design of those interfaces must be based on an intimate understanding of that work context. (p. 239)

As often mentioned in this book, the main issue is communicating information to a user with the information system being part of that communication path. The information system is a tool for connecting the user with the situation. To be a useful tool, it must ensure that only pertinent and salient information are presented and presented in the proper format and proper detail level for the user.

## Long example—Medical information system operation

This example looks at the design and operation of a system for presenting medical information to a general audience. As a way of limiting the scope, the system focuses on providing information about chronic diseases. For example, a person might use it to look up information on lupus, heart disease, or cystic fibrosis. But not to look up the new cancer research results or to find pharmacological information. Of course, even developing a decently comprehensive system as described here would be a major undertaking.

Current systems tend to be split between two groups: high level simplistic information that only gives summary answers, and highly detailed information aimed at someone with a medical background. But a vast number of users do not fall into either group. They want comprehensive information written and presented in a manner appropriate for their reading level, medical knowledge, and culture. Also, various research initiatives are starting to explore how to give patients Web access to their medical records (Wilson, 2003). Research that seems focused on security and data access issues, not on providing records designed for medical personnel in an format understandable to a layman.

Interestingly, the underlying problem addressed in this chapter, delivering information to a general audience, is much more difficult than delivering information to a medical audience. A medical audience has a relatively consistent knowledge background and a high literacy level. Through experience, it also has relatively consistent cultural expectations. None of these assumptions apply to the general audience. Literacy levels and knowledge levels cover the gamut from barely functionally literate to extensive biomedical knowledge. Cultural expectations are equally as varied. However, to provide this audience with information requires the design of a system that can adapt to and overcome this problems.

### Scenario 1

A person has just been told that her mother has Parkinson's disease and needs information about the disease. This person expects to be the primary caregiver as her mothers condition worsens.

As the person works to understand the situation, there are a large number of questions that must be answered for the person to understand Parkinson's. Many of these questions are not as obvious as they might appear, thus the system must provide the information without the person explicitly asking them. Also, some of them might not be highly relevant right now (at the point of initial diagnoses), but will be highly relevant later. For example, consider the questions that follow. Of course, these questions are much too broad for concise answers; as the person works to answer them, new more detailed questions will emerge.

- How does the disease progress?
- What is the life expectancy?
- How is the disease treated? What are the side effects?
- What help is available?
- How does the caregiver cope and handle the patient?
- What is required to care for a Parkinson's sufferer?

### Scenario 2

A person has been diagnosed with high blood pressure and high cholesterol, and told to follow a new diet that is very different from their current diet.

The situation involves gaining an understanding of the underlying problems high blood pressure and high cholesterol, of the new diet, and effects of failing to follow the treatment. Rather than sweeping statements, the information should be specific. For example, the diet information should exist at the level of specific food groups and specific types of food. It should also recognize that the person has some food allergies and not recommend them as part of the diet.

### Design characteristics

This section looks at each of the characteristics that distinguish complex situations and considers how they apply to the two scenarios using the medical information system just described.

| | |
|---|---|
| No single answer | The information required in both scenarios has no one answer. The quantity of relevant information about Parkinson's and diet changes is practical purposes lim- |

itless. Yet, the amount a user needs and wants is limited. However, how much they want depends on the particular person. Some people would want a basic "just the facts—tell me what I need to do" presentation and others want a detailed explanation of those facts and why the proscribed actions must be followed. Depending on the user, how the answers they need must be different. For example, some may need information written at a level appropriate for a high school textbook and others for a college textbook.

Open-ended questions

As part of having no single answer, the questions have no clear point at which the question can be said to be answered. Even the basic initial question "What is Parkinson's disease?" has no fixed answer. The response can run from a one sentence high level answer to a highly detailed one discussing biochemical reactions within the brain.

As the person uses the system and becomes familiar with terminology of the situation, the text can change. Essentially, the person's medical knowledge level changes and this change must be reflected in the presentation of information, especially information that was previously read.

Multidimensional strategies

The strategies of what information a person wants varies depending on how they view the situation. The person might be content with a very high level description of the disease, but wants highly detailed information about how to care for a patient. Likewise, the person looking for diet information might be more concerned with information on how to stick to the diet than on what food to eat.

Has a history

The disease is progressive and the person's information needs change over time. The system needs to capture that history to provide both what the person needs now and to modify the presentation so that is relevant with respect to the past history.

When the person first uses the system, they might read that after 2 to 8 years Parkinson's progresses to a condition of X, followed by condition Y in 1-5 more years. However, once their mother reaches that condition (let's say in 3 years), the information needs to be modified to account for how quickly or slowly the change occurred and how that tends to affect future dis-

| | |
|---|---|
| | ease progression. If condition X is reached rapidly, does that mean to expect condition Y in 1 year or in 5 years or that there seems to be no relationship between how quickly the two conditions develop. |
| Dynamic information | Besides lacking a single answer, the overall situation is dynamic. As the disease progresses or the person changes their diet, their goals and information needs and consequently the questions they ask will change. By connecting the information with the history, the system should provide information that is updated to reflect the dynamic nature of the situation. |
| Nonlinear response | The conditions that trigger a person wanting lots of information about a topic are unpredictable. As they gain a better understanding of the situation, they might want a highly detailed information about one area and decide to ignore another. |

### *System operation*

This example considers the interface operation in the gray area between the user goals and information needs and actual (underlying) system implementation.

First as disclaimer, I realize that, as drawn, Figure 6.2 would be highly unusable because the slider controls force the user into self-appraisal. Nor do I claim that an operational system would only require four sliders; I have no trouble imagining several more. A high quality operational system must have these sliders hidden from the user. Instead, it would be generating user profiles and should be dynamically adapting the presentation based on profile, user interaction pattern, and whatever else the system can collect. However, from the designer point of view, the dynamic user profiling is external to the immediate concerns of information presentation and ensuring the system supports user goals and information needs. In the end, the user profile and dynamic adaptation gets translated into virtual operation of these slider controls. Likewise, the data stored in the system requires proper metadata to allow it to be related to the sliders.

As the user moves the slider controls, the total document is dynamically changed to correspond to the slider positions. After the sliders are set, the user would be able to browse the site without touching them again. The text is a hypertext document with links to associated information. In other words, the user receives a complete (virtual) web site dynamically generated on request. Depending on the user's medical knowledge and desired detail, different readers would see different information customized for their individual goals and information needs.

Notice that nothing in this description actually describes how the text appears on the screen. The focus of this example is a high level description of a system meeting the user goals and information needs, not defining the information architecture or design. Depending on the actual information, various visualization techniques, graphic and text formats would provide the most effective presentation.

The actual presentation formats would be derived as part of the analysis. For that matter, how the information gets into the system is also outside of the immediate scope. But the final result must still be an integrated presentation and not a list of links to various web sites.

## *Slider control definitions*

Detail required — The amount of detail the user wants. This can range from basic explanations to highly detailed explanations about the underlying physical process.

For a more realistic design, the detail level has to vary between information elements. For example, the person may be content with a high level description of Parkinson's, but want very detailed information on medication side effects and patient care.

Reading level — Reading level of the user. The text needs to written at the proper level.

Medical knowledge — The medical knowledge the user has. This influences word choice and how much support information must be provided. This differs from reading level in that a person may be highly literate and can follow a complex explanation, but has minimal knowledge of medical terminology.

Cultural aspects — Cultural aspects that influence how the user sees the situation (Davis & Flannery, 2001).

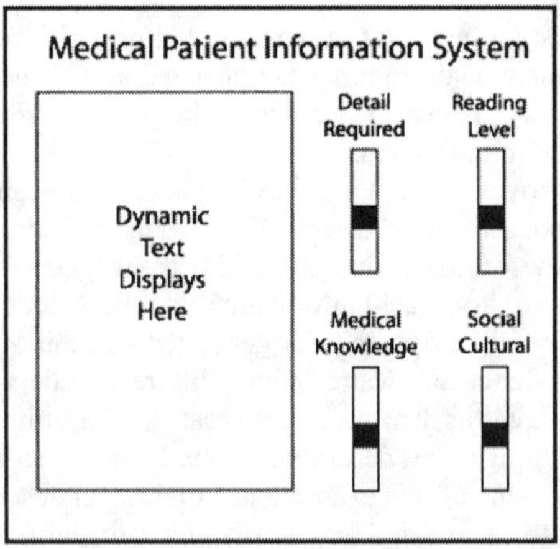

FIG. 6.2. Sketch of a possible underlying design of a medical information system. The document is dynamically updated as the sliders are positioned by the user.

# 7 Process for Addressing Complex Situations

*Design is a multi-faceted exercise. It is not a single dimension of usability or aesthetics, or cost, or time to market, or pleasure. It is all of these, and more besides.*
*—Donald Norman*

*You'll never have all the information you need to make a decision. If you did, it would be a foregone conclusion, not a decision.*
*—David Mahoney*

Each new Nielsen Alertbox article seems to set off a round of discussion on web page navigation on the Internet discussion lists about interface design and information design, yet Nielsen writes with a very strong navigational focus. Likewise in the discussions I participate in on the SIGIA-L discussion list and books such as *Information Architecture for the World Wide Web* (Rosenfeld & Morville, 1998). In the end, there seem to be many guidelines for navigation and content categorizing and few guidelines for actual content. From a somewhat cynical viewpoint, it seems most of the literature looks at content as shown in figure 7.1. A disconnect seems to exist between the need to actually communicate information and the methods of arranging all the information for a web site. The literature discusses how to create links, general page design, and how people use hypertext systems, but not how to determine what information to present. Most methods assume the actual information already exists, or it refers designers to task or user analysis to determine the information needs, effectively sidestepping the primary question of what or how information is required by the reader.

This chapter presents ideas on how to determine what content to present. It attempts to operationalize the ideas developed in the previous chapters and looks at how to implement them. It's the content area, the spot labeled "user text goes here" in Figure 7.1 that is the most important part. Yes, the information architecture is important, but content provides the information people need to achieve their goals. And only high quality integrated content provides a efficient path to understanding a complex situation. In the end, the site information architecture arises from understanding the user's goals and information needs.

Complex situations require the system to supply users with integrated information to assist in meeting their goals and understanding the overall situation as it changes. As Bodker (1991) stated, "*the design activity must be structured according to the use practice and not according to the technical components of the user interface or any other abstract or formal framework*" (p. 143). Thus, the analysis must work in terms of objectives, functions, and resources (Rasmussen, Pejtersen, & Goodstein, 1994).

For complex situations (and we've seen that most real-world situations tend to qualify as complex) the analysis should focus on uncovering the user's goals and information needs. In this process, the analysis must ensure to allow for the following characteristics of a complex situation:

- No single answer
- Open-ended questions
- Multi-dimensional strategies
- Dynamic
- Has a history
- Nonlinear response

The analysis must not focus on static needs or overly general aspects of the situation.

This chapter presents an 8-step approach to defining the users goals and information needs. This process only creates the goals and information needs; system development methodologies are still required to convert these findings into a working system. Although the steps are discussed separately, in many cases they will be achieved simultaneously and will definitely require an iterative approach for the analysis to capture all the required information. An approach which Vicente (1999a) provides a clear justification for the following:

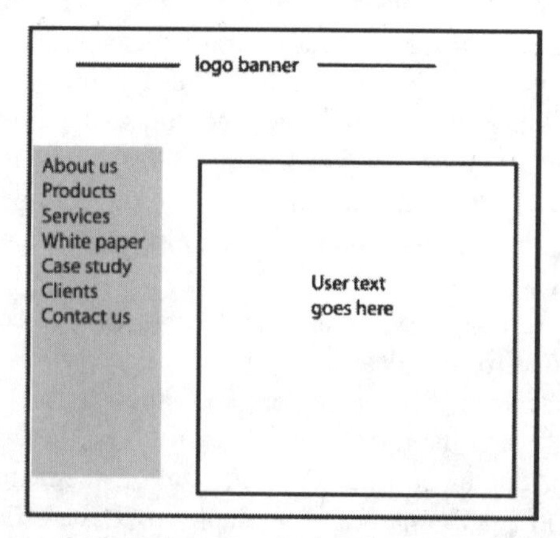


FIG. 7.1. An all too typical designer's view for a system. The area labeled "user stuff goes here" is ignored or downplayed as non-interesting. Of course, this page layout drawing must exist for any system, but the design team must not stop here. The user needs high quality information to appear in the "use stuff goes here" area as much as they need effective ways of navigating to that screen.

> Perhaps even more important, however, is the fact that it is not possible to determine the limits on our knowledge....Each iteration is a data-driven step, and we have no knowledge of the overall space of possibilities in which we are searching. (p. 106)

As Vicente points out, when we start on the analysis, we often have only a sketchy idea of what we will find. Only by performing the analysis and examining the goals information needs can we clearly understand the overall information space in which the user operates.

## Process of defining user goals and information needs

Determining the user goals and information needs and shaping the information for presentation to the user follows several basic steps. This section provides a high level overview of the steps and following sections expand on each one.

1. **Plan the scope of the information product**

   What are the complex situations for which a user needs information? The focus in this book is not overall web site architecture, but rather a method of supplying integrated information within the overall architecture. Ensure the scope statements are specific enough to be used as a guideline. Break them up into manageable units. Sometimes the user groups are too diverse to work with as a single unit (e.g., physician information and patient information).

2. **Identify the user groups of the final information product**

   Define all the potential user groups. Cooper's (1999) method of defining personas and Carroll's (1995c) scenarios provide effective methods of both defining and fleshing out the user groups.

3. **Define the user goals by user group**

   Collect the user goals for each group or user role. A coherent set of goals proves to be a major aspect that defines a user group. Because of individual needs and prior knowledge, each user within a user group may not articulate the same set of goals, but will generally need information about all goals that are available.

4. **Define the information needs of each user group or role to achieve the goals**

   Define the information required to achieve each user goal. Just as important, define the order in which the information must be presented, how it should be presented, and what makes it important to the situation and user goal.

5. **Define the information relationships that connect the information to goals**

   What makes the information relevant to the goal and how does it either influence or relate to other information? Information only makes sense with respect to other information, the relationships define that sense making.

6. **Determine the source of each information element**
   In a modern corporate environment, the information will never be in a single repository. Where is the information located and who is responsible for maintaining it?
7. **Build a visual model**
   Create a visual model that contains the overall goal structure, the information needs for the goal, the relationships. The model also captures an outline of the information content and the metadata the system will use to dynamically generate the content.
8. **Define the presentation requirements**
   Define how the information fits together to give an integrated presentation and how the user needs to receive it.

This list has carefully ignored factors that are critical to both a successful project and to a functional system, but that from the viewpoint of this book do not directly influence the definition of the user goals and information needs. Although by no means a complete list, this includes factors such as:

- Time to produce the information.
- Who will be writing and maintaining the information.
- Choice of tools for development, data storage, and presentation.
- Overall information architecture issues. This analysis only provides for the "user text goes here" from Figure 7.1.
- Implementation issues such as how to store the data and how to retrieve it.

I am not discounting the importance of these factors. A successful implementation requires them to be carefully accounted for, but they are outside the scope of this book.

The analysis should be performed in an iterative manner to allow for deeper analysis of points discovered during a particular method or cycle. The cyclic approach provides better overall usability than attempting to design the perfect system on the first attempt (Bailey, 1993) . Karat and Bennett (1991) found that highly flexible design methods work much better during early design. The cyclic nature of the data collection allows verification of the derived relationships and provides a means of gathering more information from the users. On the other hand, whereas an iterative cycle is important, too much iteration wastes time and adds little to the final results; two or three cycles probably will reach the point of diminishing returns (Hartson & Boehm-Davis, 1993).

## Plan the scope of the information product

What are the complex situation in which a user needs information? The focus on this book is not overall web site architecture, but rather a method of supplying integrated information within the overall architecture. At a minimum, the scope statements list the complex situations in which the user operates. They can also define, at a high level, the individual elements in the information architecture and type of information the user will need.

The scope definition here works in an iterative process with defining the users and user goals, the next two steps in this process. Arising out of that analysis will be an abstract template for developing the information. At this point, only the abstraction exists because neither the actual information content nor the presentation format is not defined. However, it does provide the basic outlines to use when defining the content requirements.

- Scope statements must be specific enough that they can be used as a guideline for checking the content quality.
- Define the context and situations for which the content is relevant.
- Connect the scope to the user group definitions. The scope normally is highly interconnected with the user group definitions. Often the user groups are too diverse to work with as a single unit. For example, consider physician information and patient information. The basic prior knowledge and information needs of the two groups (actually medical and non-medical groups in general) are so different that it would make no sense to try to have a single dynamically generated response for each question.
- The scope for each template should contain information that is the same type of information. The quantity of information, the detail level, and even the reading level may change between the user groups, but the basic underlying content that is being communicated should be the same.
- Break the scope of each template into manageable units. You can't answer questions about life, the universe, and everything, with one dynamically generated response. (The static response is 42, a dynamic response requires at least 3 responses.) A scope of provide information about Notre Dame (see sidebar) is too broad. Instead, focus in on specifics for which the information requirements can be defined and combined to fit the user needs.

## Scope of medieval cathedral information

Assume the site's task is to provide information about medieval cathedrals. While an overall scope statement could be just that "Medieval cathedral information" it's much too broad. Even narrowing the scope to just Notre Dame is too broad. That equates to a user saying "Give me everything about Notre Dame." What does the person what know about Notre Dame? The user's goals and information needs are typically much more refined in scope. The scope needs to focus in on the individual goals. It also needs to define aspects such as whether the information should provide comparisons between cathedrals.

**Sample scope statements**

- Rose window information (constructions, significance of images, comparison to other cathedral's rose windows).
- Stained glass windows (construction, images such as artwork, religious significance).
- Flying buttresses (construction, purpose, structural dynamics).

Scope must come from the users and fit within their user goals and information needs. The all-too-frequent methods of talking to the user's managers or being given high-level or "the corporate approved way" of addressing the situation does not provide the information needed to truly understand the reader's situational context. The focus must be on the user's vocabulary, the user's reasons using the system and the user's problems.

Remember that complex situations have no single answer. For a simple situation, the scope statements can take the form of "tell the user how to perform X." However, for complex situations, users address issues for which no ready-made answers exist in any database. Thus it cannot be expected that the relevant information be found by direct means, but inferred. Inferring answers requires comparing and contrasting information. Thus, the scope statements should also recognize that information manipulation by the user is imperative in most instances.

On page 75, the example of user goal for designing a table saw was discussed. The scope would include statement such as: "maximum cut of an 8-foot 2 x 6" rather than general statements such as "cut wood." In design meetings, when someone points out that the saw will not cut an 8-inch beam, the point can be considered irrelevant since it is outside of the scope.

## Identify the user groups of the final information product

Perform a user analysis to define all the potential user groups. User groups can be both end users (people looking at a web page) and people using that information as a partial source where they will combined the information with other information. The same people often appear in different user groups depending on how or when they use the system; an idea Constantine and Lockwood (1999) explore in their writing on usage cases. User groups are defined with respect to the situation. For example, "production manager" may fail to be a user group because situations they address in their office may be very different from the situations they address on the production floor. The final analysis may list two user groups of "production management group" to handle office-based issues and "production control group" to handle complex situation on the production floor, a group that may also include production foremen.

The fundamental nature of any user analysis must be to capture the abilities, attitudes, and aspirations of the users. Capturing these three qualities aids in the design of high quality information that meets users needs in a manner consistent with the situational context. A major objective of the user analysis is to provide the designer with the relevant data to ensure the system design matches the user view of the situational context, since forcing users to work within a system-designated method (as opposed to user-designated) interferes with understanding the situation.

During the initial stages of the user analysis strive to list all possible groups. Later, once the user goals and information needs of each group are defined, many of the groups will probably be similar enough that they can be combined.

The following list gives some factors to consider when performing the user analysis. As part of the analysis, ensure that besides capturing the relevant information about the group to capture the information that differentiates how one user groups goals and needs differ from the other user groups and how the goals and needs of a group changes as the situation evolves and becomes better understood.

| | |
|---|---|
| Information use | Define the main reasons why the user group will be using the system. This provides a foundation for understanding the type of open-ended questions and the multiple strategies the user groups will use. During later analysis, this can help clarify the detail level and presentation methods. |
| User attitudes | Define how the user group understand themselves with respect to the overall situation and their interaction with the information system. |

## User groups of medieval cathedral information

There are multiple users groups. As a first cut, I can define four. A careful analysis could break some of these down further and probably define more.

- Professional medievalists. This group can be further split into cathedral specialists and all others. The prior knowledge of the two groups probably differs radically, with the latter possibly approaching the knowledge level of the knowledgeable amateur.
- Knowledgeable amateur. A person who enjoys learning about the medieval period, but does not require or want an academic presentation.
- Student researcher. A student interested in particular aspects with enough detail to contribute to a research project, but not written at such an academic level so as to be difficult to understand, whether by writing style or assumed prior knowledge.
- Person looking for general information. This includes people who are planning a trip to Europe and want a basic understanding of what sites they will be seeing. It can also include people looking up information because they saw a TV show and want more information.

Coming out of this analysis would be three distinct groups that would require comparable information: professionals, knowledgeable amateurs and student researchers, and people who want general information. Each of these groups require substantially different amount and type of information. When fed back into the scope definition, it could reveal that three different templates would be required to address these groups.

| | |
|---|---|
| Situation control | Define how the user group controls (or does not control) the complex situation in which they will be seeking information. This can also include social or political factors that exert a strong influence on how a user group responds. |
| Group demographics | During later analysis, this can help define prior knowledge, reading levels and detail levels. It can also help provide information on the general demands of the user group with respect to satisfaction and usability. |
| Usage environment | Define the type of system access and how the user group is situated within the overall environment. A fast-paced environment or noisy environment has different presentation requirements and can be the distinguishing factor between what could appear to be the same user group. |

Cooper's (1999) method of defining personas provides one effective method of both defining and fleshing out the user groups. Having named personas for each user group lets the development team clearly discuss and analyze the user goals and information needs with respect to the groups presentation.

## Define the user goals by user group

Collect the user goals for each group or usage role. User goals are the high level view that allows the entire situation to be understood in context. User goals provide the means of categorizing and arranging the information needs.

For a simple situations, it is possible to define all user goals and enforce the order in which the user addresses those goals. For the complex situations, there is no single answer or single way of viewing the situation. Thus, each user will develop a unique set of goals, although the goals within a user group should form a coherent set. Actually, the coherent set of goals proves to be a major aspect that defines a user group. Because of individual needs and prior knowledge, each user within a user group may not articulate the same set of goals, but they will generally need information about all those goals available. As a result, a much broader collection of goals can end up being collected. The final presentation design will have to handle presenting information relevant to what goals the user has selected.

Keeping the definition of user goals and information needs separate is difficult and in most analysis situations the two will be done simultaneously. However, they serve very different purposes and operate at very different levels within the complex situations. Without the user goals, the designers risks obtaining a huge stack of information needs that lack a clear user-defined association between the information elements. However, it is important to define user goals as a way of capturing the dynamic nature of the complex situation.

With human nature being what it is, the user goals and information needs do not simply require unearthing, but require a dedicated effort to assemble them from the bits and pieces which each involved person brings to the discussion. "The best analysts have always known that settling requirements is neither a matter of simply finding out from users or clients exactly what to build nor a matter of our simply telling them what it is we are going to do for them" (Constantine & Lockwood, 1999, p. 486). Rather than simply discovering existing information or having it magically conjured up, the final view of user goals and information needs becomes dynamically created as part of the initial and ongoing analysis based on the input from a range of people. The final result is not an assembly of collected information, but a synthesis which is much greater than the original whole.

Consequently, supporting complex situations requires reconsidering many of the standard considerations of stable requirements, exhaustive task analysis, and ignorance of cognitive interaction (Rouse & Valusek, 1993). The design must provide for a deeper understanding of users, user goals, and user work processes than are common in current design (Carroll, 1995b). The overriding design task of supporting user goals becomes understanding the logical relationship between goals and available information, the reasons why they are seen as connected, and when the connection is relevant.

## *Defining and collecting user goals*

When defining the user goals, besides generating a list of goals, ensure the following list is collected. The goal here is to collect all potential goals. The actual goals set for each user group will vary while achieving the same top level goal and not all users within a user group will necessarily set all the following goals:

- Mental goals being set by the user.
- What defines that the goal is achieved. This is not the actions or information needed, but the characteristics of the situation that allow the user to declare the goal achieved. An important clarification: with information collecting common to complex situations, the user often cannot completely achieve the goal (i.e., it is impossible to know everything about Notre Dame) but will consider it achieved when the knowledge gained suffices for the current purpose.
- What defines that the goal is relevant to the situation and needs to be achieved.
- What defines that the goal is incorrect or inappropriate. In other words, the user should not be pursuing this goal.
- Importance of the goal to the situation. Some goals can qualify as a "nice to have, but not necessarily" while others must be addressed.
- Sequence in which the goals are set. Capture both the linear goal sequences where one must follow another and the parallel groupings where order of achievement is not important.
- Hierarchy of sub-goals for each goal. What sub-goals need to be achieved to declare a goal achieved.
- Relative importance of goals. Understand the logical relationship between goals and available information, the reasons why they are seen as connected, and when the connection is relevant.

- How different situation histories can affect the sequence, importance, or hierarchy of sub-goals.

During the collection of goals, start arranging them into a visual model. See the appendix for a detailed discussion.

Maximizing the information collection process and finding the users' strategies requires using multiple analysis methods (de Jong & Schellens, 1998). The following list (by no means complete) gives some potential methods that are useful for defining user goals. No one method works for all situations and often multiple methods must be applied to the same situation to obtain a complete view of the user goals. The methods listed here were previously developed for conventional task analysis. The methods work well for analysis of complex situations; the major change is the analyst must shift from thinking in terms of tasks to thinking in terms of user goals.

| | |
|---|---|
| Personas | Personas were developed by Cooper (1999) as a way of clearly defining who the users are and capturing their characteristics. Rather than working with abstract sets of lists of user attributes, personas work to develop a set of fictional, fully rounded people that work within specific situations. When design questions arise, they can be addressed by considering how the various personas would respond to that question. |
| Contextual Inquiry | Contextual inquiry provides an excellent way to studying the users in real-world situations and capturing how they work within those situations. As an ethnographic-based methodology, it basically consists of interviews and focus groups with users and, afterwards, the researchers analyze the data into affinity and workflow diagrams. It provides a very good method to define how users interacts with their environment. However, as developed by Beyer and Holzblatt (1998) contextual inquiry attempts to define the users' work processes and workflows, rather than deriving the goals and information needs behind the work processes. Thus, when applying it to complex situations, the designer must ensure the data collection focuses on goals rather than workflows (Beyer & Holzblatt, 1998; Holtzblatt & Jones, 1993). |
| Scenarios | Scenarios provide a narrative of a process or sequence of actions (not individual acts) with a viewpoint based on the users. There big advantage is focus on getting users to be explicit on what they expect, rather than the vague answers which results from more indirect methods. On the downside, unless carefully designed, sce- |

narios may not adequately cover the actual problem and the amount of information collected can overwhelm the analysis process (Carroll, 1995c). Closely related to scenarios are use cases of Constantine and Lockwood (1999).

Focus groups

Focus groups can be effective in early design work but they also can lead to group think or lead to people saying either the official corporate view or what they think the analyst wants to hear. Also, people are very prone to saying they do one thing  and actually doing other. Thus, the findings of a focus group must be verified. Often rather than focus groups, individual interviews can provide more useful data which also lends itself to comparing answers between interviews for verification.

Critical Incident

Using structured techniques like Critical Incident Technique, defined as "a set of procedures for systematically identifying behaviours that contribute to success or fail-

## User goals of medieval cathedral information

The user goals vary between user groups, but with some overlap. To keep this example manageable, only one overarching area will be discussed:  the rose window at Notre Dame. The following table lists some of the goals each group might set.

| Knowledgeable amateur | Student researcher | Potential tourist |
|---|---|---|
| Construction details.<br>Close up images.<br>Affect on population.<br>Comparison to other cathedral rose windows. | Images displayed.<br>Construction details.<br>Size information. | Images displayed.<br>Size information |

Within each of these goals sub-goals can be defined. The sub-goals exemplify the open-ended nature of complex situations. Consider the goal of understanding the construction details. Student researcher probably only need basic medieval stain glass construction details. Knowledgeable amateurs may want to go beyond that and have detailed construction information or wants to see how the rose window construction differs from the other windows in the cathedral. Or they may want to see how the different artisans assembled the glass pieces in slightly different ways.

All of the questions raised in the previous paragraph have no complete answer. Users just keep looking at information and setting new sub-goals until they are satisfied. It also shows the multi-dimensional ways people can look at the situation, depending on what they want to know about the rose window.

| | |
|---|---|
| | ure of individuals or organisations in specific situations" (Emmus, 1999). A major advantage of using a critical incident technique is that it captures the goals and information needs that occur at critical points (often major problems); however, these goals and information needs do not necessarily contribute to routine events. Many users do not go to the information systems until some sort of crisis or problem has occurred. In these situations, critical incident techniques directly apply. |
| Crystal-ball method | Helps to uncover methods of identifying informational discrepancies. Gaze into the crystal ball and see a potential situation that needs to be addressed. Then determine what goals and information needs would address the problem. As a recursive method, gaze into the crystal ball and observe what problems can occur when dealing with the situation and then figure out how to get around that problem. |

The reason that scenarios and personas are so useful is that they impart tacit knowledge of system requirements, rather than the explicit declarative knowledge captured in a more conventional process. What that means is that the specificity of the tasks and requirements captured by conventional task analysis can make it harder to understand how the system should behave when confronted with an additional requirement later in the development process. The problem is that the new requirement isn't captured in an existing task, and often the new requirement doesn't have an obvious place in the framework in the requirement structure. Personas and scenarios are about stories, narrative, and understanding underlying goals of the system. When confronted with the new requirement, personas and scenarios have created a framework for understanding that can better accommodate flexibility and adaptation The designer can fit the new requirement into the flow/plot based on an understanding of the 'stories' and 'characters' in the narrative

## Define the information needs of each user group

Using the same methods as used for defining user goals (and probably at the same time), collect the information needed to achieve and verify those goals. Work to construct a complete set of information required by the user to fully understand and evaluate a situation and to provides the interface designer a conceptual representation of the user's view of the complex situation. Also work with the understanding that although different user groups may have the same goal, the information they require may differ based on factors such as prior knowledge and how they approach the situation.

Because the complex system contains a high degree of freedom and a large amount of unpredictability, understanding information relates to the overall system is imperative to helping to address the situation. When the system work properly, the information describing the system contains normal values. The first part of the analysis must layout what these normal and expect values are. The next part lays out the incorrect or erroneous values: Don't try to figure out everything that can go wrong—a hopeless task, but define an operational definition of what it means to have incorrectly set a goal or not be adequately moving toward achieving it. Then define what is needed to help users understand which aspect is not working and why, and what information is needed to get them back on track. The final result contains the information and relationships required to ensure users can enhance their awareness of all information relevant to the current situational context.

It can be tempting to eliminate information because it doesn't exist currently, proves too hard to define, collect, too difficult to handle within the system development time frame, or, being unstructured, proves to difficult to handle within the system. There is also the temptation to ignore all information which cannot be stored or displayed by the system, information which comes from social interaction or outside sources, processes which are too often discounted (Braudes, 1991). However, ignoring this information, can compromise the effectiveness of the overall system. Accept that "anyone's real work practice is intricate and complex" (Beyer & Holtzblatt, 1998, p. 3) and involves a convoluted set of information needs. Never eliminate information as unavailable or too hard to find without thinking real hard about it.

The information collected for each goal should include:

- What information is needed to achieve the goal.
- What information is needed to verify the goal is proceeding alright or has been achieved.
- What information shows the goal has reached a critical point or a decision point.
- What information is required to show the goal is inappropriate.
- What information is irrelevant to this goal, but seems like it might be important.
- What information is required to verify or cross-check information values.
- Order in which information is needed.
- Salience (relative importance) of the information across all goal-related information.
- Amount of detail required.
- How the information should be presented (text, graphs, etc.).
- How do constraints (time, money, politics) affect the information.
- External information (not included but should be known). This information will not be stored within the system, but will be obtained outside of it. However, it still influences user interaction with the system.

### *Information needs and complex situation characteristics*

Information needs connect with a single goal. But what information a specific user wants is unknown. Thus, to provide the information must be collected with a clear understanding of how it relates to the user goal and the overall situation. Each of the characteristics of complex situation brings out the importance of this idea.

| | |
|---|---|
| No single answer | Without fixed starting and stopping points, there is no single path or sequence to follow, instead, users obtain dynamically sequenced information to help achieve a user-defined, dynamic sequence of goals. Rather than attempt to define a single path, work to define the information required for all the potential sequences and the reasons for taking a particular sequence.<br>In complex situations, people can't follow the non-existent a fixed path, instead they continually adjust their mental view and associated paths as new information presents itself. Keeping the user focused on their chosen and the relevant path means understanding how all the information elements relate to each other and to the overall situation. |
| Open-ended questions | The user questions are open-ended, so there is no fixed set of information which can be collected to answer the question. |
| Dynamic information | The information can change in value and also can be updated with respect to detail level or content. The information cannot be prestructured and must be designed to allow users to continuously adapt the information while searching for a solution (Elam & Mead, 1990). |
| Multi-dimensional | Because users approach achieving the goals differently, their information needs will be different for each approach. |
| History | The history affects how people interpret the information and how they will predict the future information values. |
| Nonlinear | The information a person wants emerges as part of the interaction with the situation. |

### *Include conflict and differing views*

Complex situations are characterized by a lack of a single answer and open-ended questions; nothing can guarantee which information user's desire or will use. Instead, the analysis must uncover potential information requirements and presentation (Blandford & Young, 1997). The analysis must not attempt to define the "correct" path; it must accept and try to provide information for multiple paths. Rather than define a path, it attempts to define what sources of information may be used regardless of how the user may approach the problem (Treu, 1992). As such, the information designer must prevent the design or development group from focusing on single paths or suppressing alternative solutions. The obvious textbook information needs were probably uncovered very early (probably before discussions with real users started). But the analysis must also uncover and lay out the implicit information relationships, social influences, assumed expert knowledge needs, and real novice questions and problems, not what experts expect novices to need.

Capturing multiple paths often results in or they arise from conflict among people within user groups. Whereas Easterbrook (1994) considers conflict inherent in requirements definition and, while many methodologies seek to avoid it, conflict must be handled and can improve understanding. Unlike many analysis methodologies, analysis for complex situations does not treat conflict resolution as an issue. Rather than avoid conflict or seek compromises, seek to exploit the conflict. For example, during a group discussion, groups often start to develop a single solution. This may happen as either a result of "following the dominant group member" or from people's desire to avoid conflict. However, strong group members must never be allowed to suppress other paths. Many methodologies discuss the need to facilitate and resolve conflict to reach the single solution (Easterbrook, 1994). This arises directly from the goal of defining the path to the solution; a reasonable goal for a well-defined task. However, for supporting complex situations, the discovery and refinement of multiple paths should be encouraged. Rather

## Information needs of medieval cathedral information

User goals vary between user groups, but with some overlap. To keep this example manageable, I'll only focus on the rose window at Notre Dame and even then, the information listed is incomplete.

Depending on the person's specific goal, the information needs can be as varied as:

- How medieval windows were constructed. This can range from basic overview to detailed description of glass making, coloring, and cutting.
- How the rose window fit into the overall cathedral concept of connecting the parishioner to heaven. Why were rose windows a common element of cathedral design?
- The type of images appearing in the rose windows.
- How the glass as survived over the centuries, including current problems with acid rain and other pollutants.
- How the Notre Dame rose window differs from other cathedral's rose windows.

than attempt to constrain users to one path per problem, which leads to trouble when the system information deviates from the user's situation needs, strive to uncover and elaborate on multiple paths.

### *Include abnormal situations*

Exploration of unusual or special conditions must be explored (Gribbons, 1991); these must be supported because they are the precise events that cause the greatest need for the system. With unusual conditions, the informal knowledge people intuitively use breaks down and then they require external support. The abnormal situation, by putting a person in a situation that violates their intuition, make people explicitly think about what information they need and how they go about monitoring the situation.

The best support for complex situations comes when the system aids in identifying discrepancies between observed and anticipated behavior of the situation (Yoon & Hammer, 1988a). The abnormal situations provide a good jumping off point for bringing out past problems that would probably not be recalled during a standard question and answer interview. Thus, the collection of information needs should cover both normal and abnormal situations; the wider the range of anomalous situations covered, the better the resulting system (Shneiderman, 1992; Gribbons, 1991).

Repeatedly, research has found most people wait until they realize a problem exists to use the system. Only then, in the search to explain unusual conditions, do most people turn to the information system (Rasmussen, Pejtersen, & Goodstein, 1994). As such, systems focused on providing information optimized for the normal situation receive low user satisfaction because it fails just when people need it most. As Carroll and Rosson (1996) clearly state the problem "A common example is the tendency to consider only scenarios of routine and expert errorless performance, overlooking the pervasiveness of user scenarios incorporating slips, mistakes, and confusions" (p. 237).

Special attention should be paid to events that interfere with clearly defining the information needs by bringing in conflicting data (Belkin, 1980). Abnormal situations to focus on include instances when normally marginal data contains information which greatly effects the outcome. Also, areas in which the outcome is very sensitive to the data values should be emphasized. Both ideas contribute to the nonlinear response of complex systems.

### *Include social interaction*

Complex situations normally involve social interactions that must be analyzed within the social setting. The dynamics are too different between observing a person alone and that same person interacting with other people to draw valid parallels. Human nature being what it is, people behave differently singularly and in groups and their behavior within a group varies depending on their status within the group. These differences can confound the analysis performed under less than realistic conditions. As Allen (1996) points out:

> As individuals, people are engaging in individual perception, alternative identification, and alternative selection on an ongoing basis; at the same time, they are being influenced in all of these activities by the social situations in which they are embedded. (p. 88)

Understanding the social interactions means the social environment must be considered when considering the information needs. Depending on the specific situation, much of the information comes directly from social interactions and will never be captured into the information system's database. Complicating social interactions are people who have their own agenda and are more anxious to promulgate it than to listen to whatever anyone else has to say. A corollary of this is that people who do listen will hear your message through their own filter of what's important, what's believable, what they already know to be true. Social relationships that often drive understanding the current situation and lets the user develop a full awareness of the factors contributing to that situation (Mirel, 1992; Gutwin & Greenberg, 1997).

The social structures must be considered a fundamental design element and the information designed to support using and maintaining the informational system within the social context (Brown, 1986). Bannon (1986) points out the biases of social structures and organizational contexts. These biases must be discovered during the early analysis and allowed for in the resulting design in order to maximize system usability. Ethnographic work should prove invaluable in identifying potential biases and translating them for application to a specific situation. Remaining cognoscente of social interactions, helps to keep the user goals and information collection viewpoint focused on a real-world context, rather than a system context (Thompson & Coney, 1995).

Of course, the social relationships vary between different user groups. The complexity of trying to account for the differences often makes it easier to simply ignore the problem, a path which can lead to a useless system. Duin and Hansen (1996) believe the social view supplies the context to make the information relevant to a real-world situation. Odell (1985) points out most people understand a situation in terms of their own immediate environment, rather in terms of the true environment. And most certainly not in terms of some generic environment defined by the analyst. Bridging the gap and developing a design which effectively supports different groups requires understanding the specific social factors that drive each group's world view.

## Define the information relationships

The relationships that exist within and between information elements and various aspects of the situation need to be defined. Information does not exist independently; it only makes sense with respect to the situation when viewed within the web of information relationships. The user will always try to form this web, the ease of doing it depends on understanding them and incorporating them into the design. The relationships define what makes the information relevant to the goal and how it either influences or relates to other information. Some examples of information relationships to define are:

- What values should track together. One action might be appropriate when two values are both increasing, while a different action is appropriate when one increases while the other increases.
- What information should be viewed together. Understanding often requires needing to compare information located in different documents and not just read the information contained within one document.
- What information is subordinate to this information. Create an information hierarchy.
- What other information is related to this element. The system must provide indications about both the existence of the other information and provide an efficient means of obtaining it (Albers, 2000).
- What information would show this value is correct or incorrect. Ease of verifying incorrect values can reduce errors and help the person identify they are using the wrong mental model.

Basden and Hibberd (1996) criticize the view that tends to believe that all the information needed can be defined in advance and then collected into a database. In this view, the knowledge exists as external to the system and user. The system simply needs to retrieve the pertinent data and present it. In this view, the data speaks for itself and information acquisition becomes little more than simply reading the facts provided by the system. This view misses the situational context and relationships inherent in information. Information never exists independently; raw data might, but never information. Data only becomes a specific type of information within a specific context. It cannot simply exist to be read, rather the reader needs to go through a process of information acquisition to transform the data into information and, ultimately, knowledge (Allen, 1996), a transformation driven by the information relationships.

Interconnected parts of the system should not be separated for the convenience of the members participating in the goal-driven analysis. A "divide & conquer" analysis philosophy is a fast-track to failure. Basden and Hibberd (1996) believe that both collecting the information and using it are knowledge-generation activities, and that often new questions arise that could not have been considered in advance.

## *Information relationships and complex situation characteristics*

Rather than attack the design problem of defining information relationships as a problem of creating a hierarchy of information, addressing the user's goals and information needs requires defining the information relationships between information elements. Each of the characteristics of complex situation brings out the importance of this idea.

No single answer — A common goal of most system design is to define a fixed path to the solution. However, in complex situations, people can't follow the non-existent a fixed path, instead they continually adjust their mental view and associated paths as new information presents itself. Keeping the user focused on their chosen and the relevant path means understanding how all the information elements relate to each other and to the overall situation.

Open-ended questions

The information relationships often drive the relevance of the information to the current situation. The information cannot be prestructured and must be designed to allow users to continuously adapt it while they work to understand the situation. The system must also try to prevent the user from ignoring relevant information or spending too much time viewing irrelevant information.

Dynamic information

The analysis must acknowledge that the relationships are themselves dynamic and subject to change. Failing to remember this leads to inherent flaws of designing to a snapshot of the overall system and the consequent failure of the resulting system. In complex situations, parts of the information can be constantly changing and the user's information needs will be changing as the situation develops. The interconnections that allow the information to be updated in an effective manner without confusing the user come from understanding the relationships.

Multi-dimensional

When users approach the situation from different viewpoints, their view of the relevant and salient relationships change. As with the dynamic aspects of the information, the system must allow users to modify the relative relationships to adjust to their needs.

## Information relationships in medieval cathedral information

Most historians are adamant that understanding history is not knowing dates, but knowing how the events of those dates influenced other events. In other words, they define understanding history as understanding information relationships. Information relationships are the basis of essay tests versus multiple choice tests.

When developing the information for a medieval cathedral, it must be more than a long list of facts: how tall, how many statues, how many mosaics. Instead, it must address how factors such as height influenced the design. How did flying buttresses let the builder's go higher. What psychological factors were the builders using when they put in large windows and many small alcoves. How did construction of one cathedral or a war influence the design of another. How did each master builder put his unique stamp on the design.

History

Complex situation analysis requires anticipating the influences of the situation history on its future. The situation history has can have a profound effect on the relationships by changing the relative importance of two relationships. Also, the users can start from slightly different beginning positions, which can affect the information relationships. A snapshot-type analysis of freezing and analyzing a static view of the situation attempts to ignore the history aspects; the resulting problem is often explained as "requirements changed." The requirements didn't change, the overall situation was never clearly understood.

Nonlinear

The nonlinear aspects of complex information means that how people value and form the information relationships change as they interact with the situation. It also means that those changes cannot be clearly predicted. The person may want more information on a particular relationship, or may want to verify if a new, previously undiscovered relationship exists.

In effective design, the design should not revolve around the information's structure, but the users' intuitive strategies and relationships. To allow construction of relationships beyond the local level, the document must contain cues that help connect the major components. An analysis that has mapped out the users goals and information needs allows the information designer to develop a system that reflects these goals and ensures that the information appears in the proper place and in the proper relationship to fit those goals. The information model must be constructed which maps these goals (Rockley, 2002).

### *Positioning goals and information relationships within the situation*

When the users complain about being unable to find the information and yet the development people working on the project can point to the information, the project is failing because it fails to allow for how people expect to find and mentally process the information. The system contains the information, but the relationships to interconnect that information with the situation do not exist. When defining the information relationships, define them in terms of the five Ws that journalists toss about.

Who

Who are the people that will be using the information? the people who provide the information content.

Observe people in their natural environment to see how they go about solving problems, performing daily tasks, identify what information resources they use and how they use them.

| | |
|---|---|
| Where | Where do they expect to find the information? Match the user's expectations, not the system's hierarchy. If the user can't find the information, then it's not in the system, regardless of whether it actually is. How do the users think the information is organized and how do they expect to find relationships? Business executives, data analysts, and clerks all take very different views of information. |
| What | What information is the user going to need? Be very specific here and describe both the content and the order the content must be provided. There may be a set of social, political, or cultural conventions and expectations that need to be addressed. The results of violating the expected conventions can range from user confusion to legal problems. |
| Why | Why is the information being looked up? What need and overall goal is motivating the user to take the time to search out the information. |
| When | When will the information be looked up? Describe the situation and circumstances exactly. Factors such as time pressure and noisy environments can make big differences in what constitutes clear information. |

The early elements of the analysis must focus on *what* needs to be accomplished and *what* information needs to be used. It does not focus on the *how* or *who* will be accomplishing the goal. The importance concept here is ensure the analysis clearly understands what goals and information needs are before progressing to more detailed design. Before designers can figure out how to support addressing a situation, they must understand what situation is being addressed.

The *what* requirement must come before the *how* because in complex situations, people can accomplish the *what* in multiple ways, which are all equally correct. Imposing a single method of accomplishing the task limits the design and, if the method does not fit the users' expected method, the overall design suffers. Perhaps even being rejected as unusable.

Eventually, the analysis must consider *who* uses the information. In a highly complex situation, multiple people can interact toward the solution. This simplistic solution makes all the information available to everyone in the same presentation. Unfortunately, this method causes overload as it includes information irrelevant to some groups. It may also run into organizational factors such as when management can access information which non-management can't. In the end, different people need different levels or types of information. Or the presentation should be dynamic and not rigidly enforce information exclusion (unless for strong organizational, ethical, or legal reasons).

## Determine the source of each information element

After the information needs are defined, the source of each information element must be defined so that it can be retrieved when the user needs it or updated as necessary. Tannenbaum (2002) points out this simple sounding step is often forgotten and "the location of data an information was never considered an issue" (p. 55).

Source information must also capture the group or writer responsible for maintaining it. For text data, the writer may have to change it. For transactional data, the programs generating the data must be recorded in case of problems and to understand when the data is posted.

In a modern environment, the information will never be in a single repository. For example, the information presented to the user could be a mix of data from all of these sources (and I have no doubt that I've missed many possible sources).

| | |
|---|---|
| XML database | Much of the static text might be stored in an XML database with an appropriate extraction engine assembling it on user request. |
| Relational database | Transactional data would be stored in a relational database. While the text might be coming from an XML database, the numbers (e.g., production or sales figures) would come from a relational database. |
| Real-time data | Real-time data captured from the situation. Not only does this need to be presented, but it must be updated on an appropriate time scale. |
| Non-controlled text | Text that is not placed into an XML database or tightly controlled in its production. For example, memos and informal status reports. |

### Information sources for medieval cathedral information

Where will the information come from? Will it all be developed and stored in a local database or will it reside on other servers around the world. How much currently only exists in books or journals and will need to be scanned or retyped?

Information on a medieval cathedral is simpler than most complex situations because the information because it is basically static. Most complex situations have dynamic information which must also be allowed for and located at run-time when the user requests it.

## Build a visual model

Construct a visual model (see appendix) of the goals, information needs, and information relationships of each user group or role. Depending on the overlap between user groups, different groups may require different models. This model is comparable to the information model discussed by Hackos (2002), although she was more focused on development of procedure-based texts. Figure 7.2 situates the model within the overall development process. This analysis gives:

| | |
|---|---|
| What information we need | Why they need it |
| How the information maps to goals | When they need |
| The information relationships | Who needs it |
| What we have | How much the know |
| What we need to develop | How to tag for retrieval |
| How they need it (level and detail) | Format for presentation |

The modeling process for most topics may be too complex to place on one sheet of paper or even in one large diagram. For example, the overall goal and subgoal structures can get lost if all the information requirements and relationships are connected to them. Use a large wall and draw the goal, information needs, and relationships. The modeling process is similar to that used in other visual models such as workflow diagram (Hackos & Redish, 1998), topic maps (www.topicmaps.org) or affinity diagrams (Affinity Diagram; Gaffney, 1999). After building a visual model, you will have the following:

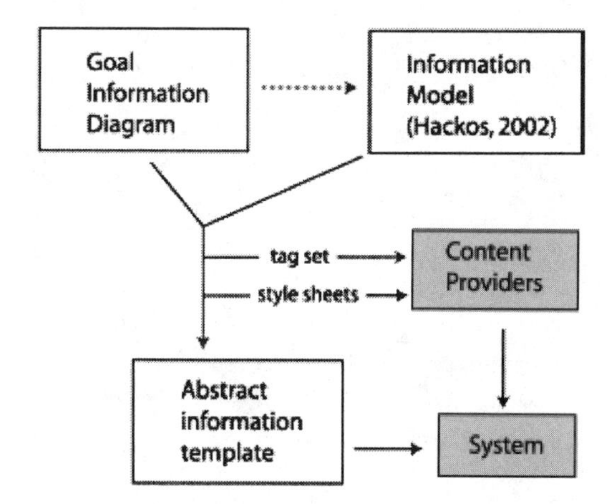


FIG. 7.2. Location of the visual model within the analysis process. Information needs from the goal info diagram is used for the information model analysis of the existing data.

- Visual representation of the high level structure of how the goals relate to various user groups.
- The information needs connected to a goal.
- Outline of the content of each information element.
- List of locations of the required information.
- Visual picture of the relationships between information elements.
- Visual diagram of how the information can be dynamically combined to suit the user needs.
- Visual picture of how the user sees the goal and information needs (the user mental model).
- Metadata required to connect the information through the relationships.

### *Define information elements and content requirements*

The analysis thus far has defined the information needs, but has not considered in detail how they information fits together. Now, as part of developing the information model, how the information needs interconnect and what information is actually required to communicate that information need must be defined.

Defining the required content of information units themselves are a major goal of the analysis to support complex situations (Albers, 2002). Developing an outline of the size and content of the information elements is an exercise in synthesizing the audience analysis, the user goals and information needs, and the system technology limitations (Banfalvi, Sturgeon, & Walsh, 1996). Once the contents, the relationships, and the presentation of the information units are known, then the job of supporting complex situations is well along.

## Visual model for medieval cathedral information

A visual model helps to clearly lay out what information should be presented for various required detail levels. Missing on this drawing, but important for a more realistic work, would be a coding scheme to support different knowledge levels.,

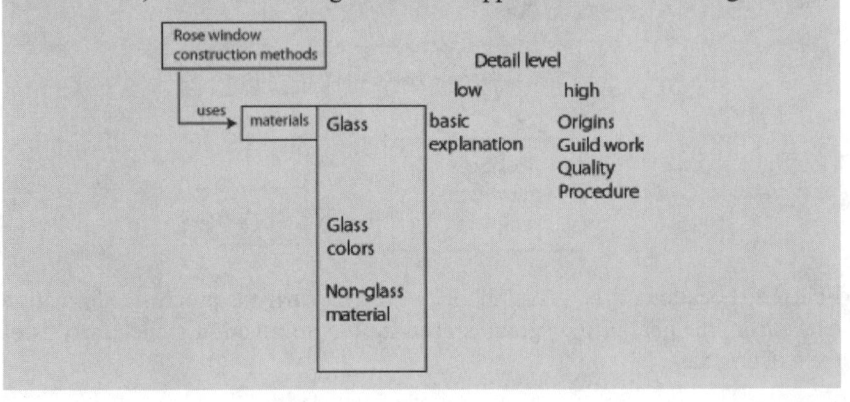

The information element must be designed to make sense as a stand alone object and to allow combining with other information elements for an integrated presentation. Each information element needs to be self-contained, written to contain all the information the user needs but nothing more. In general, it should conform to the mantra given out they the single sourcing advocates (Hackos, 2002; Rockley, 2002). But then, the dynamic information systems discussed here are a form of single sourcing and most of the single sourcing literature applies directly.

Creating a coherent information that can be combined into dynamic document means ensuring the information elements are self-contained and encapsulate the pertinent information. In paper documentation and static web pages, the static nature of the design makes it easy to define a self-contained information unit. On the other hand, with dynamic information, an efficient implementation requires information reuse and, by definition, construction of the information at run-time.

With dynamic information, the definition of both complete and self-contained take on different meanings because of ability to dynamically add or subtract information based on user input. Complete and self-contained exist at multiple levels, all which must be defined within the goal and information model. A set of relationships that drive the usefulness of the information for different users. Of course, different usage roles require different views of the same information which include different detail levels. Each one requires a different level of content to comprise the information element.

An information element exists as an interface construct and not an underlying implementation construct. There is no reason for the entire information element to exist as a single piece of text. As such, it has no requirement (and probably not desirable) to exist as a unified element in the physical system structure. By their nature, the overall contents of an information unit may be defined in the initial analysis, but the actual contents is dynamically built when the information is needed, just before display. The information element itself may consist of a mix of both text and pointers to text that will be merged when the user wants to view it.

### *Define the metadata required to relate the information to user goals and user groups*

Connected to the information elements are the metadata which contains the relationship information. The metadata allows the system to dynamically generate the content in a way that creates a coherent presentation, rather than as a collected of disjoint information elements. As the Rockley Group (2002) discussed in a white paper:

> Metadata can be used to describe the behavior, processes, rules and structure of the data, not just descriptive information. These elements are important when developing a sound metadata strategy for content search and retrieval, and dynamic content delivery, because they determine not only what the content is, but who uses it, how it will be used, how it will be delivered, and when.

The metadata provides a means of labeling and cataloguing the information, a way of showing the relationships, and the rules governing the information use and relationships. The metadata structure should arise from the information structure and the limitations of the specific tools planned for implementing the system. For instance, in an XML implementation, the metadata would take the form of tag attributes.

### *Define the mental models of each user group*

Using the goals and information needs, construct a visual map of the basic mental model for each user group.

The analysis must draw out the different mental models held by people with different views of the problem or different experience levels. Collecting these mental models requires working with real users, a technique that adds perspectives and insights that a designer working alone would miss (Lanning, 1991). User preconceptions can have a strong affect on the acceptance and usability of a system. These preconceptions must be molded into an explicit form during the goal-driven analysis (Terwilliger & Polson, 1997). If the novice and expert mental models differ radically, the goals and information needs of both must be collected or the system resulting from the analysis risks being useless to one or both groups.

The properly designed system must transform the system model into the user's model and not force the user to think in terms of the system model. When the match occurs, the structure of the presentation matches the user's mental model and enhances effective system use. As a result, a user can easily acquire the required information to enhance the understanding of the situation. Without a clear conception of the user's mental model, the design process risks getting drawn into the system level details while missing critical elements for meeting the user's mental model. (Note that the concern here is with the presentation of information; communication occurs between the user interface and the person. The structure and implementation of the underlying system design is irrelevant, except as it supports the structure and implementation of the presentation and communication of information to the reader.)

The map of the mental model held by a user group provides help at many stages of the design and implementation.

- Basis for resolving design questions about how to present information.
- Foundation for analyzing design changes or additions to determine the potential affects of the change.
- Method of keeping the user's big picture view in front of the development team.
- QA checkpoint for comparing the model against the actual system design to ensure the interface and presentation conform to that model.
- Input for analyzing user problems or misunderstanding with the presented information

## Define the presentation requirements

At the beginning of the chapter was Figure 7.1 with "user text goes here." This step begins the process of deciding what goes there and how it goes there. The focus here is not on creating the information itself, but on deciding what information elements will go together and how they interact with other information elements.

Enhancing users' understanding of the situation requires information which clear presents information supports the goals and gives an understanding of the problem and the possible solutions. Supporting user performance means structuring the analysis and resulting design to reflect the user practices and not the technical aspects of the implementation (Bodker, 1991).

Effectively connecting the user goals and information needs to an interface can be the hardest part. Hard because a computer screen is too limited and impoverished to effectively support the highly parallel processing human. A small display area (even large monitors have small display areas compared to the total information volume) complicates the communication process, taxes user memory, and leads to information overload (Johnson, 1997).

When designing the presentation requirements, allow for an overall design that is a small web site and not a single web page. Remember to consider the following:

- How do the information units built in the previous step fit together to give an integrated presentation.
- How can the information relationships be maintained and clearly presented.
- How does the user view the information hierarchy.

### Information presentation for medieval cathedral information

More than just the basic web page architecture, the presentation contains other factors such as how much animation, photos, audio, and text to use.

How should these presentation elements be combined for each audience? Depending on audience needs, they might vary.

- A highly knowledgeable audience may prefer a primarily text-based presentation supported with photographs. A person looking for general information might prefer a short overview with lots of photos and a short animation of how the window was constructed.
- The amount of term definition and technical terminology varies with the knowledge level of the audience.
- The amount of detail a person wants is independent of their knowledge but should be adjusted to fit whether they want a highly detailed presentation or just a couple of photos.

Effectively handling the last two bulleted items requires careful construction of the metadata to allow for dynamic information construction that fits multiple user profiles.

- How can the user be cued to information not currently visible.
- What is the most salient information and how can it prevent getting lost in large amounts of other information.
- How can the dynamic nature of the text be constructed to meet the user's needs.
- How can the information best be presented: video clip, graphics, text only?.

Any non-trivial system has a high level of complexity, but the design must hide it (Eden, 1988). The designer must capture and understand the complexity with respect to the business and reader viewpoints to have a starting place for hiding the complexity for the user and to clearly communicate the information. Draw on the user's mental models to match them with the presentation requirements. Accessibility based on organizational devices (headings, table of contents, index, etc.) is more important (Goodwin, Huston, & Southard, 1988). The need for transparent information organization re-emphasizes the need to follow effective layout principles: formatting of lists and headings, visual cues, lots of white space, generous leading, readable fonts, and so on. All contribute to the user acceptance of the information and the overall quality of the resulting system (Hartley & Trueman, 1985; Kostelnick & Roberts, 1998; Lay, 1989)..

As part of this, the template structures into which the information will be places will start to emerge. Also at this point, the entire process starts to become highly tool dependent as the information moves into databases and the design begins to reflect tool limitations. In general, the actual implementation would involve extracting the proper XML content, populating the template, modifying it based on conditions within the XML, and then binding it with a XSL schema to set the display characteristics.

## Conclusion

Conventional task analysis assumes structured tasks, but complex situations and the related problem solving occur in an unstructured environment. The methods presented here are designed to address getting a handle on the unstructured nature of the user goals and information needs.

As such, the analysis for complex situations attempts to define the user goals, what information is needed, and what limitations are imposed on applying that information and ensuring the information remains in context and preserving the interrelationships. The analysis must define both the information needs and interrelationships to ensure activation of the proper mental model and provide the information needed to flesh out the specifics of that model. Rather than attempting to define the right way, it lets people apply their expertise and ingenuity to the current situation to create an acceptable solution. It accepts that people are adaptive and will adapt to the situation. To be effective in supporting that adaptation, the model developed through the analysis must represent the user's view of the entire complex situation and not a computer system and software application.

Although methods presented in this chapter and the conventional task analysis methods may appear similar, the difference between the task analyses for structured and unstructured tasks is more than just one of minor modifications. Rather, the difference is as deeply fundamental as the overall goals of the data collection. The fundamental difference arise from the different goals of the user. In the highly structured environment, the user's basic goal is essentially one of efficiently completing the task and the task analysis consists of defining the actions and information needed to perform a task. Such production-oriented tasks have fixed starting and stopping points. However, in a complex situation, the user's goal consist of some combination of understanding, problem-solving and decision-making. Goals occur in cyclic sequences, which continues until the user feels confident with the information gained or quits in frustration.

## Long example—Online textbook design

A well-designed online textbook allows a more integrated and adaptable presentation of the subject when compared to a printed text book. This should increase student learning.

> Indeed, Kaput asserted that many of the traditionally "hard" topics of science and mathematics may become more learnable within the rich and integrated approaches afforded by combinations of highly visual, student-controllable simulations, meaningful physical and social settings that tap into students' naturally occurring linguistic, kinesthetic, perceptual and cognitive sense-making powers, and local learning objectives that build on their interests and felt needs. (Jacobsen, et al., 1998)

The dynamic information presentation for complex situation ideally fits a textbook. Both because of the open-ended nature of the learning task and of the wide range of student abilities. By providing custom and dynamic information, students can receive information suited to their individual learning style and current knowledge level in the course. When the student first reads a section, it can contain detailed explanations of the topic and later, looking at the section for reference, it can be reformatted as a reference document. Reading level of the text can also be adjusted to fit a student's needs.

Extra or more detailed material can be available for students with a high prior knowledge level. The extra material would also make the online text better for using with two courses with different audiences, such as a technical communication service course and a technical communication course limited to majors. In the latter case, the students need more underlying theory or a deeper understanding of the topics.

### *Complex situation characteristics*

Teaching is a psychological process that forms a highly complex situation. This section considers each of the characteristics of a complex situation.

No single answer

A class should have clear objectives about what the student needs to learn. Thus, it may seem that a single answer does exist. However, most of the topics could contain much more information than a student is able or willing to learn. Working various problems in the text or other assignments brings the student to the text with a slightly different view. The dynamic nature lets the book refocus to fit the student's current knowledge level and the detail required for the current question.

Open-ended questions

Besides covering a wide range, the questions are posed as open-ended questions. If the homework assignment requires a strong cognitive aspect rather than problems which require little more than finding answers in the text, then the information needs of the student vary. Most want enough information to help with the problem, but how much depends on the situation.

Multidimensional approach

How the person views the information depends on their individual circumstances and causes them to approach the information differently. Different learning styles can require different presentation. Also, a student looking back into a previous chapter for help has a different view and understanding of the information than does a student reading the text for first time.

Dynamic information

The person's goals and information needs are highly dynamic. As the student progresses through a course, their knowledge level is rapidly changing on a day-to-day basis. In a teaching situation, the goal is to increase the student's knowledge from the current level to a predefined level. Of course, current levels vary depending on how well previous work was understood and prior knowledge. A dynamic system can adjust to handle these differences and help maximize student learning.

History

The past interactions with the information need to be considered. For example, a student that is returning to a section multiple times may need it restructured so they can better understand it. Also, the previous text they looked at can influence what they are looking for, such as when doing a problem requires looking at three or four different pages to assemble the information to form an answer.

| | |
|---|---|
| | Prior knowledge in the topic, for example work experience, can influence how the student views the information and how much they need to view. Prior knowledge that is wrong or in conflict with the text can require a more deep explanation so the student understands why their understanding of the concept needs to change. |
| Nonlinear | As a person learns more about the topic, they come up with more questions. As they develop a well-formed mental model of the information and its relationships and begin to comprehend the topic, they shift from needing general information to specific points to fill in gaps; a shift which can occur abruptly. |

### *Designer and writer issues*

Chapter 7 described a set of steps to consider when designing for a complex situation. This final section looks at those steps as they apply to an online textbook.

#### Scope

The scope of the textbook matches the limits of the course. Rather than being the specific course objectives (these are user goals), the scope describes the higher level layout of the subject matter landscape and what is considered applicable to the textbook and what is not. These scope statements provide boundaries for developing information that conforms to the user goals.

#### User groups

Any class contains multiple user groups. The differences can be more pronounced for service courses and introductory class where there are students with different backgrounds. Different user groups could be based on major, prior knowledge, attitude toward the class (the required but unpopular class), learning style, and other factors that cause people to have different views and attitudes toward the material.

#### User goals

The user goals for an online textbook would closely map to the course and unit objectives. Coupled with goals based on course objectives would be user goals that support the text problems and assignments, and possible post-class use as a reference document.

#### Information needs

A combination of user groups and user goals drive the information needs. How much information each user group needs to learn the objective and how it should be presented drives the information needs.

### Information relationships

Understanding a subject comes when the person sees new information with respect to the topic's body of knowledge and not as individual factoids. Because a major objective of any class is to teach the student how to make those connections, the text must clearly lay out the relationships and make them easy to comprehend.

### Source

All of the information exists in various forms on the Web, but most of those formats are not easy to understand or mentally integrate. Students can read the individual articles, but don't have the ability to integrate them. So, the source probably does require the textbook authors to initially generate most of the information. Other pertain sources can also be found and most be clearly documented, so given the volatile nature of web pages, they can be tracked for any changes.

### Presentation

Balancing the information needs against the user groups is major difficulty in online textbook design. The traditional view uses mostly text supported with a few graphics. However, that assumes a particular student learning style. The overall design and the dynamic variations must be chosen to suit the user groups and effectively communicate the information.

# Conclusion 8

*We see what we look for, not what we look at.*
*—George Neisser*

*Complexity is the curse of the digital age. It is a type of intellectual pollution that smothers clear thought. Complexity is not a sign of intelligence, but rather a sign of a hyperactive mind gouging or more. True genius and great design is about turning something complex into a product that is simple to use and delivers a real benefit to the consumer.*
*—Gerry McGovern*

Current single sourcing literature discusses taking pieces of text and assembling them into multiple documents and multiple output formats. However, it assumes that the output basically matches the documentation we have now. As has been evident throughout this book, I'm not discussing new ways of obtaining what we have now, but new ways of getting someplace else. The dynamic information that I've discussed throughout this book is not what technical communicators were taught in school, nor is it they writing the practice on the job. I realized this way back when I began writing this book. It may be closer to what cutting edge researchers in information retrieval and adaptive hypertext want to develop out of their work, but they seem to assume the content exists. I'm writing for the person who must create the content. Not the most glamorous job in the world and one that may take an even more unglamorous twist. The unglamorous part is the piecemeal development of content that will be required. A situation I (Albers, 2000) examined from the editing viewpoint and which Weiss (1993; 2002) has described as "contributing to the database stew" (p. 61) is that individual textual elements will need to be created and stored in a way that allows their eventual placement into many documents.

Why the database stew? Because that's the only way the information can be assembled to match a user's expectations, goals, and information needs. The entire purpose of the model I've developed is to help directly match those expectations, goals, and information needs within specific situations. The old way of creating documents, each writer working on a single static document, lead to the final output of text that only gave information one way. But similar information should be communicated in a multitude of ways to fit the user's goals and needs.

The audience requires information tailored along three dimensions: prior knowledge, detail level desired, and ability to understand the material. A static document must pick one spot with respect to those three dimensions and write for it; often a document that aims at the lowest common denominator on the assumption that higher level users can gain what they need. With the ability of adaptive hypertext and single source databases, there is no longer a need to be locked into picking one spot. Each audience can get the information customized for their needs. Bist (1996) talks about how information needs are changing and that:

- "Information will be continually assembled in new and unpredictable ways.
- Hierarchically structured libraries of information will give way to flatter design to accommodate 'mixing and matching' information in ways not anticipated in the original design.
- New libraries of information will be created from selected portions of existing libraries created independent of each other" (p. 49).

The goal of this book was to help figure out how to create information that will be assembled in new ways. In a somewhat ironic twist, the difficulty with designing good informational systems arises because the analysis and design is itself a highly complex problem. Vicente (1999a) sums up the problem by saying:

> We regularly encounter colleagues who have collected a great deal of data but have then found it very difficult to derive design implications from those data. They become overwhelmed in details, and find it hard to see the forest for the trees (p. 134).

The need for highly integrated information arises because real-world situations are complex situations that possess a high level of complexity with many explicit and implicit relationships. Relationships that must be understood to allow for clear communication of information.

## Goals and information needs

A major business goal of any informational system is to enhance user productivity. Accomplishing a productivity increase in complex situations means it must address the complexity of open-ended tasks and provide the methods to accomplish them.

When faced with designing systems to address a highly dynamic situational context, the need to communicate information clearly to a user do not change. The basic goal of the design is still to meet the user's goals and information needs. However, the methods of accomplishing the task of how to define the information requirements and presentation change. Gone are the ability to fully define information needs or order of presentation, or to gain a full understanding of the user's goals. In their place is the need to provide the a means of accessing the desired information in a user-defined order while still maintaining a coherent view of the information interrelationships required to support the fundamental user goals and information needs.

Hollnagel (1993) supports these ideas of defining dynamic information when he defines three goals for successfully supporting complex situations.

1. Present the right information.
2. Present the information in the right way.
3. Present the information at the right time.

Hollnagel envisions a system that provides only high-quality, current information which addresses the users' real-world questions. While three statements themselves seem obvious and trivial, a major design implementation difficulty lies in defining the user's goals and information needs, especially when wants and needs differ. Hollnagel's questions may be trivial, but clear answers are most definitely nontrivial. As conceptualized, the system sounds wonderful, but implementing the system and defining the required real-world information unleashes a whirlwind of devilish details. Yet, without working through the whirlwind and gaining the understanding of the implications of the data and how it applies within the situational context, the ability to support Hollnagel's three statements doesn't exist (Beyer & Holtzblatt, 1998).

Each of Hollnagel's three statements deserve a closer look with respect to the ideas about the design of complex information laid out in this book. In their simplicity, they provide way to sum up the main ideas as they state the overall goal.

### *Present the right information*

Although this book spent three chapters (2-4) directly focused on defining the right information, it could not provide a single format. There is no single format to help user's solve open-ended problems because there is not one answer; rather the answer to the open-ended questions depends on users' learning style and ability, their knowledge of the task, how much information they want, the type of task, and the type of documentation expected.

Good analysis methodologies allows the designer and writer to enter the user's mind and define the open-ended questions the user may ask. Only by understanding what open-ended questions might be asked can the writer answer them in a manner compatible with the user.

Almost by definition, systems the address complex situations come with high price tags for both the programming and content generation. The information design must ensure that content gets communicated to the end user to justify the cost. Designers must understand the nuances of how a design supports the text and develop the most effective way to communicate that message so that it does not sink without a trace or get lost in extraneous noise.

### *Present the information in the right way*

Achieving the goal of meeting users' goals and information needs means presenting information when and how they wants it. Meeting users' goals and information needs can require slightly different information or information in a slightly different form. Perhaps someone wants a graph and another wants a table. Or the detail level of the text needs to change. Once the goals, information needs, and relationships are defined for each user group, then the designer can figure out what does constitute the right way for that group.

The analysis should expose the multiple levels of complexity and then must draw out these levels to make them explicit so that they can be addressed during the system design phase. By working from an analysis that identifies the levels of complexity of both the user goals and needs and the information itself, the design phases of implementation can focus on hiding the complexity (Eden, 1988), but not the information, from the users.

Rather than being something the user focuses on, the content and the design must convey the message in a clear manner without interfering with that transmission. As the design begins to be noticed, the amount of information conveyed decreases because the cognitive resources devoted to handling the design (often interface design issues) aspects, rather than information, increases. Reality stops short of completely meeting the when and how aspects but the data required to relate and clarify other data should be at most one mouse click away. And the mouse click must be an obvious one; the reader must always be clear on what information sits behind which link.

### Present the information at the right time

People have a hard time handling information that arrives at the wrong time. Just as important as getting in the right way, it also needs to arrive at the right time. It can be neither too soon nor too late. Too soon and the person either forgets it or must go back and find it again; too late, and the information might as well have not existed. It will either be ignored or can cause problems as a user tries to undo a decision to make use of the new, but late, information. Woods and Roth (1988) address the issue of providing proper information by saying:

> The critical question is not to show that the problem solver possesses domain knowledge, but rather that more stringent criterion that situation relevant knowledge is accessible under the conditions in which the task is performed (p. 420).

Software can have an undo feature and machine assemble can be undone; both put you back at the exact spot you were once at and allow you to move forward along a new path. But with people-based systems, there is no undo. Any corrections or adjustments are made to the current state which has been influenced by all previous decisions and actions. Also, people have trouble extracting and synthesizing information that is separated by time or space. Right way and right time presentation means that all relevant information gets presented in a manner that the person can easily synthesize and relate to the current situation.

Post-hoc analysis of system development has many stories of systems which went unused or were used in ways quite different from initially envisioned. People will take over technology and turn it to their own uses (Dourish, 1999; Mirel, 1998). The social changes which occur as a system is implemented rapidly take on a life of their own. As with so many other people-oriented processes, these are fundamentally dynamic and nonlinear with the actually path or final outcome impossible to predict. The communication path cannot be clearly laid out, but the general route must be maintained.

## Problems in a billing and invoicing system

I once worked for a company that was designing a new billing and invoicing system which one of the lead programmers described as user-friendly because it was a GUI interface which used a mouse. If we would have been designing that system today, I have no doubt that the company would have thought simply designing it as a web application would make it user-friendly. In reality, it was the most user-hostile interface I've ever seen; one which was so bad that once the customers saw it, the plug was pulled on the project. That interface was primarily a data-entry and report printing interface; complex-problem solving wasn't even part of the system requirements. Yet, it shows how hard any interface is to design. Entering the prices of over 1000 products for many dealers, pricing the product and creating invoices, tracking customer relation problems and dealer information presents a complex situation. A complex situation that was never acknowledged in the system design.

The problem was that software interface didn't hide the complexity or help the user perform their actions. Rather it blatantly displayed the complexity with a deep hierarchical menu structure based on the system design requirements. The designers never tried to figure out exactly what problems the user were trying to solve and how they went about solving them in the current system. Instead, they took the problems, grouped them functionally, and created a menu structure. Rather than adjusting the system in the user model and context, it forced the user to adjust to the system model.

One telling example was the ease with which programmer could query records and the difficulty of a typical user doing the same task. (While the users understood their job in detail, they were mainly women with a high school education who had never been exposed to the formal logic structures required for complex queries). When a new screen appeared, all the fields were blank. Boolean queries could be typed into as many fields as desired and the system would retrieve all records matching those queries. Of course, the users had no concept of Boolean expressions and nothing indicated that the field would scroll to allow more characters than the visible width. The fix was for me, the technical writer, to put a section into the manual explaining queries and Boolean logic. I also knew the company's intent was to send one manual to each site which would have 6-20 terminals for this system.

## Open research questions and looking to the future

Researchers and the trade press point out how ubiquitous computing will soon allow the user to be always be connected. Coupled with the rapid changes in presentation technology, the need for dynamic information continues to increase. Ignoring the delivery technology and only looking at the presentation methods, we can see that the computer display on which the information appears is changing rapidly. Until recently, designers only had to worry about desktop monitors and could be confident that they would get bigger and increase in resolutions. Now the trend has radically reversed and with Palm, Windows CE devices, and cell phones starting to increase in their share of the web browser market. These tiny screens with limited graphic capabilities stand as the anti-thesis to the highly graphical desktop monitor. Also on the horizon are head-mounted display devices that project a virtual monitor into the user's eye, showing promise of a return the large display areas.

Business needs dictate faster and more global access to information and easier/quicker development and maintenance. Many writers, such as Fayyad and Uthurusamy (1996), believe current predictions of information needs and access methods underestimate reality. Criticizing future projections as extrapolations based on paper books, they consider the projections of information needs and access requirements as overly conservative. Whether the future proves Fayyad and Uthurusamy correct, goals and information needs will remain central to understanding complex situations and the amount of information available to users will continue to increase. All of which supports Marchionini's (1995) statement that:

> Although it is too soon to say how electronic environments have affected general information-seeking patterns, the early indications are that the way we think about and react to information seeking is changing as more of these systems become available to us (p. 99).

But yet, as the trade press hypes ubiquitous computing and the various delivery protocols, they hype a technology and how to push data. The content issues are ignored and it's up to the content provider to ensure the user receives something useful. It's not the protocol, the software, or the hardware that will give an answer to a person's question. Only the information content presented in the proper format can answer the question.

People will continue to be increasingly exposed to situations in which they need varying amounts of information in order to quickly and efficiently comprehend the situation. Providing that information depend on the careful analysis of the user's goals and information needs and the work of a technical communicator to ensue the information exists in a form that the system can assemble into a coherent answer.

Compared to the technology side, the writing and information design side is behind the curve in creating new ways of creating and delivering content. As I pointed out in chapter 1, there is a lack of research into the issues of analyzing, designing, and developing information for complex situations. Also, the technology people have the easy job, they work with precise logical systems to build bigger and faster pipes to move data. Content still moves from person to person. The technology is easy compared to the human psychology issues of mentally processing and comprehending information and deciding how to apply it to a situation.

The needed research covers the entire spectrum of all six characteristics of complex situations and how people operate within them. Each of those characteristics still require extensive research. Some of the main issues of what we still need to know about them are discussed.

## No single answer

Currently, most document and web design assumes a single answer exists and the emphasis is to help the user find it. Reflecting this bias, most of the research occurring in web-based systems seems to narrowly focus on navigation issues (Larson & Czerwinski, 1998). The attitude often seems to be: once the information is placed into the proper category, the problem is solved. Yes, navigation is an issue that must be understood, but, as mentioned many times in this book, great navigation is useless without real content sitting at the end of that navigation path.

Recent aspects of cognitive engineering have looked at the definition and development of an information space (Benyon, 1998; Ingwersen, 1994). How well people work within the information space directly relates to the ease of information processing. Reworking the analysis to enable the information designer to think in terms of people working within an information space should increase the power of the analysis to address highly complex problems and to help define the various dimensions of the problem. We need more work on how to define and perform information analysis of an information space.

The lack of a single answer arises in the complexity of the situational context. As a start, Clarke and Cockton (1998) have worked on development tools that support building relationships for information context within the design and development environment. But few other researchers seems have examined how to connect context with information within web-based systems. Granted, connections between context and information are situation-specific. However, broader-based research that examines multiple varied instances should reveal the underlying principals that help users mentally form the connections of context to information needs (Treu,1992). Without a layer of underlying principals, any new development project must construct a system from ground zero. With a set of principals providing a framework, the analysis could focus on relating the principals to the specific situation and ensuring coverage of the important aspects of each principal.

## Open-ended questions

Providing answers to open-ended questions puts the designer in the position of having both too much information and no single clear way of presenting it. Consider the problem with most search engines: too many hits. The person has no way of easily sorting the hits. Relevancy indicators are problematic at best as most people have no idea how to interpret them. Usable dynamic complex systems need a way of sorting information—high quality from low/wrong information and putting it into proper buckets. The large quantities of information in an enterprise scale system would require some level of automated metatagging coupled with human interaction to give guidance to the automation.

Then, after tagging and storing the information, ways of retrieving it and creating rhetorically coherent documents is required (Albers, 2000). Information retrieval has gotten very efficient at the retrieval problem. The rhetorical coherent aspects seem to have been ignored. Yet, that drives how a person understands the information and, consequently, the situation.

Another area of future research is developing tools which support analysis for complex situations and the subsequent system design (Santoni, Furtado, & Francois, 1997). Sumner, Bonnardel, & Kallak (1997) consider that systems supporting design work must "effectively support designers' cognitive work without hindering their creative process" (p. 83), a quotation which accurately sums up the primary objectives of supporting complex situations.

## *Multidimensional approach*

Handling the multidimensional factors inherent in real-world situations creates a complex research environment with an abundance of confounding factors. Developing a practical analysis methodology requires further research into the multidimensional aspects of information as it relates to both people understanding information and designers creating it. We need a better understanding of how people mentally develop the multidimensional view and how to best present information which supports it. Methods of incorporating multidimensionality into other cognitive and task-analysis methodologies to maximize the final result needs further study (Steinberg & Gitomer, 1993; So et al., 1997). Accomplishing this multidimensional research in a web-based environment requires detailed studies of how people acquire and subsequently process information from those systems.

Analysis for complex situations should use cognitive engineering as its foundation (Woods & Roth, 1988; Bowie, 1996) coupled with the ideas found in situation awareness research (Endsley, 1995). Although all these researchers recognize the multidimensionality of information in the real world, minimal empirical research has applied the concepts to complex information situations (Ingwersen, 1994). Thinking in terms of assisting the user to develop a sufficient awareness of the situational context may prove to be the element which helps transform information systems from collections of data to integrated and focused information sources. But, as far as I have found, my dissertation (Albers, 1999) was the first work to directly attempt to apply situation awareness as an explicit concept to information design. Endsley (1995), as well as Randel and Pugh (1996), has developed situation awareness measurement tools for use in naturalist decision-making, but they worked with short-term situations. Work still needs to be done to apply the concepts to dynamic, web-based information environments. Their measurement tools need to be examined and modified to fit with type of understanding and decision-making required within that environment.

### Dynamic information

Developing systems that will produce dynamic documents and reconceptualizing how to develop the text which goes into them are two areas which require extensive research (Maglio & Matlock, 1999; Card, Mackinlay, & Shneiderman, 1999).

Content management is moving from a model where content is inert and acted upon to a model where the application that acts on content and the content itself merges. When content becomes can, on request, replicate, automate, enhance and distribute itself and interact with other contents units of content to produce compound, complex information, then the "Content Ecotecture" will have arrived (Shao, 2001).

But the dynamic content brings with it its own set of problems. People have trouble with information sets that seem to randomly change. Also, it becomes an open question of how to evaluate the system and perform usability testing. With dynamic information, counting clicks or timing search time can be problematically since different information may be displayed for each user. And always using the same information (a reasonable laboratory control method) can mean that some percentage to the participants are looking at a set information which is not appropriate for them. Perhaps the testing will have to shift to comprehension tests, such as those used in reading research or freeze tests used in situation awareness research (i.e., freeze/blank the screen and ask the participant to give the situational status).

### History

Situation history coupled with the multidimensional aspect shows how understanding depends both on how the situation got to its current state and the viewpoint taken by the user. Most design and analysis comes from a view that incorrect work can be undone. But with people involved, an "undo" function does not exist. People cannot just forget incorrect answers or information; it will continue to influence their view and comprehension.

However, history factors and how they affect situation understanding have been generally ignored. There may be some cognitive psychology research into the effects of changing information and how it influences decisions, but this research has not been taken to the point that it can be easily applied to designing and writing content. We need to develop research into understanding how history affects a person's view of a situation, how to present that history, and how to capture it.

### Nonlinear

The nonlinear aspects of complex situation brings with it:

- An extreme sensitive to initial starting conditions,
- A nonlinear scaling of situation complexity. As the situation becomes more complex, the difficulty of understanding it increases much faster than the situation complexity.
- Emergent properties which can only be discovered as the situation develops and cause it to exhibit properties that are not found in the individual elements. The overall situation behavior is more than the sum of the individual components.

All of the issues have been essentially ignored in the current research literature, yet they often are a major cause of problems with systems. How to handle them and address their dynamic nature requires a focused interdisciplinary research program.

## Final comments

The analysis methods discussed in this book strive to define the user problem domain in terms of goals and the information required to achieve those goals. It seeks to build an abstract model of the factors required to maximize user understanding and performance (Hix and Hartson, 1993). Going beyond just defining the information needs, the analysis addresses how information is integrated or combined in the process of gaining an understanding of the situation. The following list sums up some of the major aspects of the viewpoint I've presented.

- The analysis is determined by the user's goals or objectives and the information required to achieve them, not conventional tasks (Hackos & Redish, 1998; Carroll, 1995c).
- The goal is to determine what the decision maker would ideally like to know, but acknowledges that incomplete information is the norm and complete information can never be provided (Blandford & Young, 1996).
- The analysis helps users conceptualize and visualize the information relationships (Treu, 1990).
- The analysis attempts to accurately represent real-world conditions (Macaulay et al., 1990).
- The analysis provides support for task-inherent, task-supportive, task-enhancing, and task-peripheral conditions and requirements (Treu, Mullins, & Adams, 1989).
- The analysis does not define a single "best" answer or path to the solution, but attempts to uncover all the common paths and provide the information needed to support them (Hallgren, 1997).
- The analysis does not assign priority to goals or sub-goals; short-term and long-term goal hierarchies should be distinct. It is assumed that during the decision-making process, the relative priority of goals change as the decision makers' understanding of the situation changes (Endsley, 1995).
- Error identification, error correction and out-of-the-norm situations are emphasized since they are the ones which require the most problem-solving support. When solving a problem, people must recognize what is wrong, what information is needed to verify the problem, and possible solutions.
- The analysis includes both static and dynamic information. Examples of static information are procedures, policies, or rules (laws) affecting the problem, information which, for the purposes of the decision, can be considered fixed. Dynamic information is sensitive to system parameters and varies on a short time scale, for example, production/sales information.

The lessons to be learned—or relearned—from this long bullet list are that:

> We need to work closely with and listen to users to understand the true nature of their work and the problems they face. Then we need to reflect that understanding in the systems we design and build. We need to treat users as intelligent problem-solvers by producing innately flexible designs that accommodate to varied styles of work and problem solving, by creating interfaces that do not enforce a rigid logic or artificial order and that do not deprive users of the opportunity to devise better ways to achieve their own ends (Constantine, 2002).

I believe this work is the first comprehensive treatment on analysis and design which considers the highly dynamic situational context of information, the aspects of the information, and the information interrelationships required to support the fundamental user wants and needs. I sincerely hope it will not be the last. The issues raised are too important for the overall goal of communicating information to a user. As I've just pointed out, too many the holes that remain in how to communicate information.

In the preface, I stated that this book will not offer a prescriptive method. I hope by now, that it's obvious why I feel there is no prescriptive method. There is only a high mountain range before us that requires an updated toolset of analysis methods if we are going to conquer those mountains and provide better, more assessable information. The user does not need a static set of information; the user needs a malleable set of information that conforms to their goals and information needs. It must be customized by their knowledge, detail level, and ability.

No one method can ever accomplish that for a complex situation. More than presenting a method, I desired to ensure the readers of this book understand that complex information situations exist, how to recognize them, and how to begin to analyze the user goals and information needs.

# Appendix: Creating a Visual Model

One of the steps of the process laid out in Chapter 7 was to create a visual model. This appendix provides more detail on what to incorporate into that diagram.

The diagram provides of a graphic representation of the goals and information needs of users for addressing complex situations. It visually displays the relationships between the users' goals and the information needed to achieve those goals. As a graphic method, it ties naturally into handling the hypertextual nature of online informational systems (Yazici, Muthuswamy & Vila, 1994). The fundamental design goal of the diagram is to assist in explicitly visualizing the relationship and cueing information. As part of its development, the full interconnectedness of users' views of goals and information relationships emerge.

The diagram should provide a clear picture of the information relationships and the user's mental model of the system which the designer can use. Coupled with the social factors found through methods such as ethnography, it then can provide a sound foundation for design decisions made during the various phases of system design and implementation. By providing a visual representation that corresponds to a user's mental model, it provides the designer with answers to these type of questions:

- How do the users define their goals?
- How do the users see a goal in relation to other goals? What do users see as a logical sequence of goals and how do they see parallel goal processes (set of goals that must be accomplished, but are independent of order).
- What information addresses the goal?
- How do users compartmentalize the information?
- How do they label the information?
- How do they see the information as interrelated?

In general design characteristics, diagrams are related to data flow diagrams, affinity drawings (Affinity Diagram), task nets (Hollnagel, 1993), information models (Hackos, 2002; Rockley, 2002) but provide a stronger focus on goal-information relationships any of these three.

By design, it should be an abstraction of the user's goals and information needs. Rather than a design intended to map directly onto an interface, but provides clear input into the interface design and information contents elements during system construction. The abstract nature of the diagram allows a designer to think large, and to explore the avenues of possible information needs while deferring details about interface widgets and other detailed design minutia (Constantine & Lockwood, 1999).

Creating the diagram consists of the four steps listed below, performed in a highly iterative manner. Details on performing each step will be explained later.

1. Define goals/problems.
2. Define sub-goals.
3. Define information needs.
4. Define cause-effect relationships between goals and goals, and goals and information.

Some factors to remember when creating a visual model.

- No categories. Having categories results in sorting notes, not finding new relationships or expanding what we know
- Represent the real information collection and interaction, not the ideal as defined by the organizational hierarchy (Beyer & Holtzblatt, 1998). The method of interaction may important to obtaining the information (email Jane, walk down the hall, call someone who is three time zones away).
- Ban common (technical) words of the development group. All terms must be user-based to keep focused on the real problem.
- Show incorrect goals of common wrong mental models. For these, the information needs would include the proper information to inform the user of why the goal is incorrect and, possibly, to help shift the user's mental model.
- Contains multiple paths to understanding the situation. The potential for multiple paths for achieving a goal should be explored and developed as much as possible to maximize the users' associative and task-oriented strategies for finding information (Heba, 1997).
- Capture the range of possibilities the user may expect from the system. The diagram is not restricted to a single or best set of goals. Users do not always follow optimal paths; the diagram should capture the information needs of these sub-optimal paths.
- Accommodate strategy shifts. User strategies tend to shift frequently in order to focus on resolving localized goals (Rasmussen, Pejtersen, & Goodstein, 1994).

- Map the approaches used by both inexperienced and experienced users. The mental model differs with experience and the ones in use at various stages of a user's experience with the situation affects how they approach understanding the situation.
- Include ways that social relationships affect understanding the situation and setting goals.
- Built on an integrated goals and information needs of all the people from the analysis. It is not a best fit with some goals and information ignored because only a subset of people want it.

## *Reasons for a visual model*

For complex situations, the dynamic information requirements exemplifies Heba's (1997) comments about how, historically, writers have been the creators of documents, but now users are becoming the creators and the writer is but a provider of possibilities. The diagram displays a set of possible paths to solving a problem and brings out users' needs and reflects those needs in terms of user knowledge and desires. The user-knowledge focus allows for designing a user-centered system that accurately reflects the user view rather than the designer view. Since it also brings out the way various user groups address a complex situation, it also helps refute any attempts at formulating an idealized strategy to support a single path for understanding a situation.

The best support for a complex situation comes when the system either integrates observations with actual behavior, or aids in identifying discrepancies between observed and anticipated behavior (Yoon & Hammer, 1988a). At least one of the user's goals or sub-goals will be the currently active goal (Endsley, 1995). The design of the system must exploit how the user moves from goal to goal and consider questions such as:

- What is a logical sequence of goals?
- What specific details cause a user to choose the next goal?
- How does this goal connect with the situation?
- What reveals this goal is not appropriate right now?

Analysis of the diagram can also provide indications of the areas most likely to act as cognitive bottlenecks to understanding the situation or as problem areas on collecting the required information either from the situation (for dynamic information) or about the situation (for static information). Based on this analysis, the designers could focus the prototyping and usability testing onto areas of potential trouble (Miller-Jacobs, 1991).

Developing a diagram for each user group helps to highlight both the similarities between them and bring out the differences. It can be especially helpful in clarifying the different information detail and knowledge levels required to address the same goal as set by different user groups. This can later help the content developers understand the level of granularity required to support all user groups and to priorities what information to focus on.

### *Analysis points to include in the visual model*

A diagram helps to clarify the entire range of information relevant to the situation and tries to make explicit four types of task knowledge: task-inherent, task-supportive, task-enhancing, and task-peripheral (Treu, Mullins, & Adams, 1989). All four of these are needed because the usefulness of an informational system depends upon the explicit knowledge and explanatory capability built into the system from the beginning (Benaroch, 1996). Defining the relationships within the diagram maximizes the users' associative and task-oriented strategies for finding information to support solving the problem. Points to identify during the analysis:

- The information directly related to achieving the goal, including the proper granularity level of the information.
- Sub-goals required to achieve the goal.
- Information needed to establish a full understanding of the situation.
- How the information is interconnected.
- The related information used to test the validity and reliability of information. For example, if numbers on two different reports should go up/down in sync with each other then it should be easy to check if that is happening.
- Information to verify any decisions made in the course of accomplishing the desired goals. For example, what/how values will change and when if the solution is moving in the right direction? What/how values will change if the solution is not moving in the right direction?
- The information which may constrain possible solutions. For example, company policy or relationships with other products.
- Variations in the information which may require setting another goal. For example, high or low report figures and possible causes of the high or low values.
- Problems which may be encountered (Belkin, 1980).
- Cues to indicate the solution is working/the proper solution has been chosen (Jones, Farquhar, & Surry, 1995).
- Information to help verify the identify proper goal is being addressed (Gribbons, 1991).
- Rhetorical techniques that maximize information presentation.
- Identification of salient information. What information is most important and what is of secondary importance? Clearly identifying the most important can help prevent information overload by forcing the user to find pertinent information in a sea of less relevant information.
- Cues to ensure proper schema activation (Orlikowski & Gash, 1994; Thuring, Hannemann, & Haake, 1995).
- Related information that might influence understanding the overall situation. For example, current or future plans for the project.

- Context for the goal and its information needs.
- Sensitivity to other elements within the situation or work environment. These may be either social or technical.

## *Construction of the visual model*

The development of the diagram consists of four steps.

1. Define goals/problems.
2. Define sub-goals.
3. Define information needs.
4. Define cause-effect relationships between goals and goals, and goals and information.

While these steps are presented as if they proceed in sequential order, in practice they merge into an iterative, interwoven web of definition and refinement.

Develop the diagram using either large sheets of paper or by hanging cards from a large wall. Like most techniques for large diagrams, use of a large wall and sticky notes works well. The creation method must allow for this dynamic change. The diagram for a complex situation will be both large and highly dynamic as it develops. Also, expect the diagram to require to be reshuffled multiple times to allow space for additional goals and information and to adjust for findings from prototyping or other user-centered design methods which feed it data.

A comment on arranging the diagram. A linear sequence often is not best; a spoke and hub model often works better, as Hackos (2002) has found when she develops information models:

> Instead of hierarchy or time line, we have moved now, in many instances, to a hub and spoke arrangement to solve a number of information- reporting problems. The hub of the information model represents the user's most immediate focus of attention – whatever key information takes precedence at a particular moment in the analysis or interrelationships. The spokes represent related information that may be viewed, discarded, viewed again, and perhaps moved to the hub position as the interrelationships change.

### Step 1: Define goals/problems

The goals users want to achieve or the problems that must be addressed are defined. These are the high-level goals or user questions that initially bring the users to the informational system. They should be expressed in user terminology and not system terminology.

Plot the goals in sequence that generally fits the order in which the user groups attempt to understand the situation.

### Step 2: Define sub-goals

Any sub-goals that must be achieved in accomplishing the main goal are defined. Sub-goals may in turn also contain sub-goals. Sub-goals may be duplicated under multiple higher-level goals.

The goals should be arranged in a spoke and hub or a hierarchical structure to the extent that they support the respective structure. In other words, the actual result normally may be many small hierarchy diagrams, rather than a single large diagram. Goals which don't interconnect should never appear within the same hierarchy and a relationship should not be forced where one does not exist.

Discovery of multiple paths to solutions should result in the sub-goals breaking into distinct independent grouping, with each group corresponding to a different path. The different paths should be annotated as such so that later design work does not attempt to combine the paths or intermix their elements.

### Step 3: Define information needs

The information required to achieve each terminal sub-goal is defined. Some sub-goals may have qualifying information that define whether the sub-goal is applicable. Information needs may be duplicated under multiple sub-goals.

The information needs should not be viewed as simply a textual entity. Example information needs could be:

- A corporate form
- Instructions for completing the form
- Samples of previous work that can be used for comparison
- Tabular reports
- Competitive evaluations from external sources
- External information that the system can only direct the user to check out. Too many of these will make for a problematical system, since the user will tend to ignore most of them.
- Discussion of the relative merits between different approaches to a problem
- Discussion of past conflicts or "gotcha" type occurrences that occurred in previous attempts to address the problem.
- Discussion of other problems with similar symptoms and how to distinguish between the various problem.
- Training tutorials for learning how to perform the task.
- Potential environmental conflicts
- Corporate policy statements that relate to the problem
- Overview information that places the problem in the larger scheme

### Step 4: Define the relationships between the elements

The inter-relationship between differing goals, goals and information needs, and information needs should be indicated. Defining the relationship as a verb phrase helps to put the two information items into context (Bolton et al., 1994; "Interrelations Digraphs" 1997). In the analysis, the user groups define the relationships rather than attempting to develop them automatically from factors such as the "term weights" commonly used information retrieval and adaptive hypertext research.

The following list gives some example relationships, but this listing is by no means exhaustive. Specific circumstances require the potential cause-effect relationships to be modified either by putting forth a fixed list or by expanding on possible choices.

| | |
|---|---|
| is supported by | inherits from |
| is undermined by | joins |
| may be specialized by | leads to |
| must be considered before | maintains |
| is a component of | is opposite of |
| is constrained by | is a portion of |
| is a constraint on | disproves |
| is controlled by | cannot occur with |
| must occur with | |

During the layout design, the information relationships on the diagram become the basis for defining what information goes on a particular page and link definition. Unfortunately for the designer, the knowledge levels of the users strongly affect how the user view links. Knowledge of the subject correlates highly with moving through the information space in a nonlinear way; the approach used by experienced and inexperienced instructors could significantly different in this situation. Thus, providing a highly intelligible link gives users quick access to that information, but, more importantly, reminds the users that the information exists. The link acts as a cue to consider the information before the user even sees it.

## Sample diagram

This section contains a sample diagram, severely limited to fit onto one page. Both goals and information needs are boxed, with the user goals in bold and the information needs in normal type. The relationship information appears in italics and sits beside the line connecting various boxes. The arrows point to the higher level of the design.

While the sample is abbreviated, it contains all the elements found in a much larger diagram.

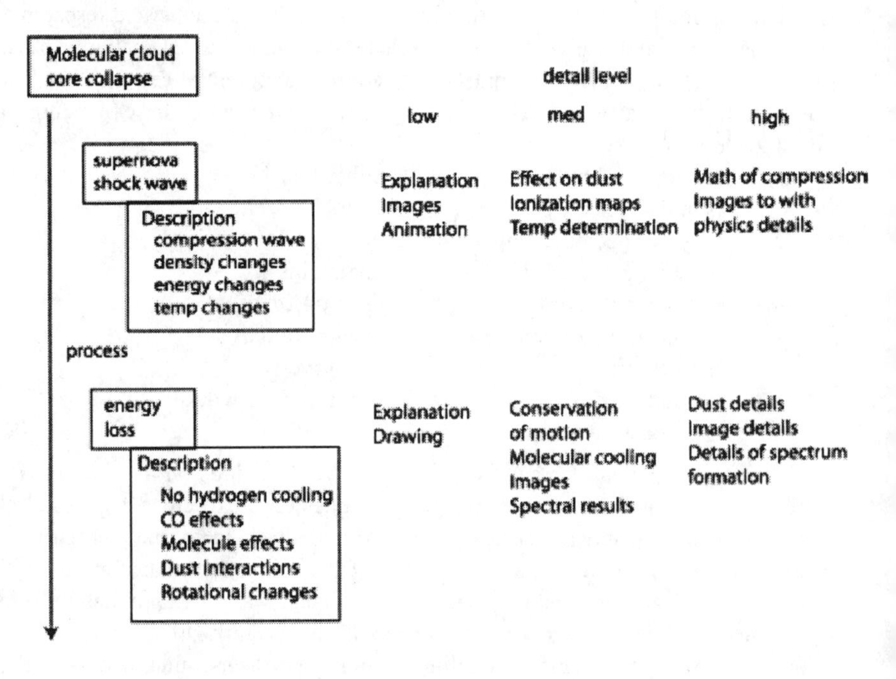


FIG. A.1. Partial visual model of a molecular cloud collapse.

# References

Access Travel Guides. http://www.harpercollins.com/hc/aboutus/imprints/access.asp

Affinity Diagram. http://www.sytsma.com/tqmtools/affin.html. February 26, 2003.

Albers, M. (2000). The technical editor and document databases: What the future may hold. *Technical Communication Quarterly, 9.2,* 191-206.

Albers, M. (2003a). Complex problem solving and content analysis. In M. Albers & B. Mazur (Eds.), *Content and Complexity: Information Design in Software Development and Documentation* (pp. 263-284). Mahwah, NJ: Lawrence Erlbaum Associates.

Albers, M. (2003b). Multidimensional audience analysis for dynamic information. *Journal of Technical Writing and Communication, 33.3,* 263-279.

Albers, M. (2000). Information design for web sites which support complex decision making. *Proceedings of the Society for Technical Communication 47th Annual Conference.* Washington, DC: STC.

Albers, M. (1997). The key for effective documentation: Answer the user's real question. *Proceedings of the Society for Technical Communication 44th Annual Conference.* Washington, DC: STC.

Albers, M. (1999). *Development of a goal-driven methodology for requirements definition in hypertext information systems supporting complex problem-solving*. Unpublished dissertation. Texas Tech University.

Albers, M. (1996). Decision-Making: A missing facet of effective documentation. *Proceedings of the 14th Annual International Conference on Computer Documentation*. Washington, DC: ACM.

Albers, M. (1997). Information engineering: Creating an integrated interface. In M. Smith, G. Salvendy & R. Koubek (Eds.), *Proceedings of the 7th International Conference on Human-Computer Interaction* (pp. 213-216). New York: Elsevier.

Allen, B. (1996). *Information tasks: Toward a user-centered approach to information systems*. San Diego, CA: Academic Press.

Anderson, R. C., & Pearson, P. (1984). A schema-theoretic view of basic processes in reading comprehension. In P. Pearson (Eds.), *Handbook Of Reading Research* (pp. 255-291). New York: Longman.

Andriole, S., & Adelman, L. (1995). *Cognitive System Engineering for User –Computer Interface Design, Prototyping, and Evaluation.* Mahwah, NJ: Lawrence Erlbaum Associates.

Archer, N., Head, M., & Yuan, Y. (1996). Patterns of Information search for decision-making: The Effects of Information abstraction. *International Journal of Human-Computer Studies, 45,* 599-616.

Bailey, G. (1993). Iterative methodology and designer training in human-computer interface design. *Proceedings of INTERCHI'93*. Washington, DC: ACM.

Bailey, R. (1989). *Human Performance Engineering*. New York: Prentice-Hall.

Banfalvi, T., Sturgeon, P., & Walsh, C. (1996). Manufacturing documentation in the virtual warehouse. *Proceedings of the 14th Annual International Conference on Computer Documentation.* New York: ACM, 161-166.

Bannon, L. (1986). Issues in design: Some notes. In D. Norman & S. Draper (Eds.), *User Centered System Design: New Perspectives on Human-computer Interaction* (pp. 25-30). Mahwah, NJ: Lawrence Erlbaum Associates.

Basden, A., & Hibberd, P. (1996). User interface issues raised. by knowledge refinement. *International Journal of Human-Computer Studies, 45,* 135-155.

Basden, A., Brown, A., & Tetlow, S. (1996). Design of a user interface for a knowledge refinement tool. *International Journal of Human-Computer Studies, 45,* 157-183.

Belkin, N. (1980). Anomalous states of knowledge as a basis for information retrieval. *The Canadian Journal of Information Science, 5,* 133-143.

Benaroch, M. (1996). Roles of design knowledge in knowledge-based systems. *International Journal of Human-Computer Studies, 44,* 689-721.

Benbasat, I., Dexter, A., & Todd, P. (1986). An experimental program investigating color-enhanced. and graphical information presentation: An Integration of Findings. *Communications of the ACM, 29.11,* 1094-1105.

Benyon, D. (1992). Task analysis and system design: The discipline of data. *Interacting with Computers, 4,* 246-259.

Benyon, D. (1993). Adaptive systems: A solution to usability problems. *User Modeling and User Adapted. Interaction, 3,* 65-87.

Benyon, D. (1998). Cognitive Engineering as the development of information spaces. *Ergonomics, 41.2,* 153-155.

Beyer, H., & Holtzblatt, K. (1998). *Contextual design: Defining customer-centered systems.* San Francisco: Morgan-Kaufmann.

Bist, G. (1996). Applying the object-oriented. model to technical information. *IEEE Transactions on Professional Communication, 39.1,* 49-57.

Black. A. (1990). Visible planning on paper and on screen: The impact of working medium on decision-making by novice graphic designers. *Behaviour and Information Technology, 9,* 283-296.

Blandford, A., & Duke, D. (1997). Integrating user and computer system concerns in the design of interactive systems. *International Journal of Human-Computer Studies,* 46 653-679.

Blandford, A. & Barnard, P. (1995). Using interaction framework to guide the design of interactive systems. *International Journal of Human-Computer Studies, 43,* 101-130.

Blandford, A., & Young, R. (1996 November). Specifying user knowledge for the design of interactive systems. *Software Engineering Journal,* 323-333.

Bodker, S. (1991). *Through the Interface: A Human Activity Design Approach to Computer Interface Design.* Mahwah, NJ: Lawrence Erlbaum Associates.

Bolton, D., Jones, S., Till, D., Furber, D., & Green, S. (1994). Using domain knowledge in requirements capture and formal specification construction. In M. Jirotka & J. Goguen. (Eds.), *Requirements Engineering: Social and Technical Issues* (pp. 141-162). San Diego: Academic Press.

Borenstein, N. (1991). *Programming as if People Mattered: Friendly Programs, Software Engineering, and Other Noble Delusions.* Princeton, NJ: Princeton UP.

Bowie, J. (1996). Information engineering: Communicating with technology. *Intercom, 43.5,* 6-9.

Boyle, C., & Encarnacion, A. (1994). Metadoc: An adaptive hypertext reading system. *User Modeling and User Adapted Interaction, 4,* 1-19.

Braudes, R. (1991). Conceptual modeling: A look at system-level user interface issues. In J. Karat (Ed.), *Taking Software Design Seriously: Practical Techniques for Human-Computer Interaction Design* (pp. 195-207). San Diego: Academic Press.

Brehmer, B. (1988). Organization for decision making in complex systems. In L. Goodstein, H. Anderson & S. Olsen (Eds.), *Tasks, Errors, and Mental Models* (pp. 116-127). Philadelphia: Taylor and Francis.

Brown, J. (1986). From cognitive to social ergonomics and beyond. In D. Norman & S. Draper (Eds.), *User Centered System Design: New Perspectives on Human-computer Interaction* (pp. 457-486). Mahwah, NJ: Lawrence Erlbaum Associates.

Brusilovsky, P. (1996). Methods and techniques of adaptive hypermedia. *User Modeling and User-Adapted. Interaction, 6,* 87-129.

Butler, K., Esposito, C., & Klawitter, D. (1997). Designing more deeper: Integrating task analysis, process simulation, and object definition. *DIS '97.* Amsterdam, The Netherlands: ACM.

Campbell, K. (1995). *Coherence, Continuity, and Cohesion.* Mahwah, NJ: Lawrence Erlbaum Associates.

Card, S., Mackinlay, J., & Shneiderman, B. (Eds.). (1999). *Readings in Information Visualization: Using Vision to Think.* San Francisco: Morgan Koufmann.

Carliner, S. (2000). A three-part framework for information design. *Technical Communication, 49,* 561-576.

Carlson, R., Wenger, J., & Sullivan, M. (1993). Coordinating information from perception and working memory. *Journal of Experimental Psychology: Human Perception and Performance, 19.3,* 531-548.

Carroll, J., & Rosson, M. (1996). Getting around the task-artifact cycle: How to make claims and design by scenario. In M. Rudisell, C. Lewis, P. Polson & T. McKary (Eds.), *Human-Computer Interface Design: Success Stories, Emerging Methods, and Real-World Context* (pp. 229-268). San Francisco: Morgan-Kaufman.

Carroll, J. (1995a). Introduction: The scenario perspective on system development. In J. Carroll (Ed.), *Scenario-based. Design: Envisioning Work and Technology in System Development* (pp. 1-18). New York: Wiley.

Carroll, J. (1995b). Work processes: Scenarios as a preliminary vocabulary. In J. Carroll (Ed.), *Scenario-based. Design: Envisioning Work and Technology in System Development* (pp. 19-36). New York: Wiley.

Carroll, J. (Ed.). (1995c). *Scenario-based. design: Envisioning work and technology in system development.* New York: Wiley.

Casaday, G. (1991). Balance. In J. Karat (Ed.), *Taking software design seriously: Practical techniques for human-computer interaction design* (pp. 45-62). San Diego: Academic Press.

Castel, F. (2002). Ontological computing. *Communications of the ACM, 45.2.* 29-30.

Chalmers, M. (1999). Informatics, architecture and language. In A. Munro, K. Hook & D. Denyon (Eds.), *Social Navigation of Information Space* (pp. 55-79). London: Springer.

Chase, W., & Simon, H. (1973). Perception in chess. *Cognitive Psychology,* 4, 55-81.

Cilliers P. (1998). *Complexity and Postmodernism: Understanding Complex Systems.* Routledge.

Clarke, S., & Cockton, G. (1998). Linking between multiple points in design documents. *Adjunct Proceedings of CHI'98.* Washington, DC: ACM.

Cohen, M. (1993). The bottom line: Aiding naturalistic decision. In G. Klein, J. Orasanu, R. Calderwood & C. Zsambok (Eds.), *Decision Making in Action: Models and Methods* (pp. 263-278) Norwood, NJ: Ablex.

Cohill, A. (1991). Information architecture and the design process. In J. Karat (Ed.), *Taking Software Design Seriously: Practical Techniques for Human-Computer Interaction Design* (pp. 95-114). San Diego: Academic Press.

Collette, G. (1991). Knowledge management: The future direction of technical communication. *Proceedings of the 1991 STC Conference.* ET-45-48.

Conklin, J. (2003). Wicked problems and fragmentation. http://www.cognexus.org/id26.htm March 7, 2003.

Constantine, L., & Lockwood, L. (1999). *Software for Use: A Practical Guide to the Models and Methods of Usage-Centered Design.* New York: ACM Press.

Constantine, L. (1999). What do users want? Engineering usability into software. http://www.foruse.com/Files/Papers/whatusers.pdf.

Constantine, L. (2002). The Emperor has no clothes: Naked Objects meet the interface. http://www.foruse.com/articles/nakedobjects.htm Dec 10, 2002.

Cooper, A. (1996). Three models of computer software. *Technical Communication, 43.5,* (229-236).

Cooper, A., & Reimann, R. (2003). *About face: The essentials of user interface design.* New York: Wiley.

Cooper, A. (1999). *The inmates are running the asylum.* Indianapolis: Sams.

Cypher, A. (1986). The structure of users' activities. In D. Norman & S. Draper (Eds.), *User Centered System Design: New Perspectives on Human-computer Interaction* (pp. 243-264). Mahwah, NJ: Lawrence Erlbaum Associates.

Dattolo, A., & Loia, V. (1996). Collaborative version control in agent -based. hypertext environment. *Information Systems, 20,* 337-359.

Davis, R., & Flannery, D. (2001). Designing health information delivery systems for Puerto Rican women. *Health Education & Behavior, 28.6,* 680-695.

De Jong, M., & Schellens, P. (1998). Focus groups or Individual interviews? A comparison of text evaluation approaches. *Technical Communication, 45.1,* 77-88.

Desouza, K. (2003). Facilitating tacit knowledge exchange. *Communications of the ACM, 46.6,* 85-88.

Diaper, D. (1997). HCI and requirements engineering—Integrating HCI and software engineering requirements analysis. *SIGCHI Bulletin 29.1,* http://www.acm.org/sigchi/bulletin/1997.1/diaper.html May 5, 1997.

Doheny-Farina, S. (1988). *Effective Documentation: What We Have Learned. From Research.* Cambridge, MA: MIT.

Dourish, P. (1999). Where the footprints lead: Tracking down other roles for social navigation. In A. Munro, K. Hook & D. Denyon (Eds.), *Social Navigation of Information Space* (pp. 15-34). London: Springer.

Duffy, T. (1995). Designing tools to aid technical editors: A needs analysis. *Technical Communication, 42.2,* 262-277.

Duin, A., & Hansen, C. (1996). Setting a sociotechnological agenda in nonacademic writing. In. A. Duin & C. Hansen (Eds.), *Nonacademic Writing: Social Theory and Technology* (pp. 1-15) Mahwah, N J: Lawrence Erlbaum Associates.

Duin, A. (1991). Reading to learn and do. *Proceedings of 38th STC Conference,*

Dye, K. (1988). When is document accurate and complete. *IEEE Transactions on Professional Communications, 31.2.*

Easterbrook, S. (1994). Resolving requirements conflicts with computer-supported. negotiation. In M. Jirotka & J. Goguen (Eds.), *Requirements Engineering: Social and Technical Issues* (pp. 41-65). San Diego: Academic Press.

Ebert, D., Zwa, A., Miller, E., Shaw, C., & Roberts, D. (1997). Two-handed volumetric document corpus management. *IEEE Computer Graphics and Applications, 17.4,* 60-62.

Eden, C. (1988). Cognitive mapping. *European Journal of Operational Research, 36,* 1-13.

Eggleston, R. (1997). Adaptive interfaces as an approach to human-machine cooperation. In M. Smith, G. Salvendy & R. Koubek (Eds.), *Proceedings of the 7th International Conference on Human-Computer Interaction* (pp. 495-500). New York: Elsevier.

Einstein, G., McDaniel, M., Owen, P., & Cote, N. (1990). Encoding and recall of texts: The importance of material appropriate processing. *Journal of Memory and Language, 29,* 566-581.

Elam, J., & Mead, M. (1990). Can software influence creativity? *Information Systems Research, 1.1,* 1-10.

Emmus. (1999). Critical incident technique. http://www.emmus.org/html/frames/guidelines/EmmusWP3/methods/cit.htm

Endsley, M. (1995). Toward a theory of situation awareness in dynamic systems. *Human Factors, 37.1,* 32-64.1

Fawcett, H., Ferdinand, S., & Rockley, A. (1991). Organizing information. *Proceedings of 38th STC Conference.*

Fayyad, U., & Uthurusamy, R. (1996). Data mining and knowledge discovery in databases. *Communications of the ACM, 39.11,* 24-26.

Fennema, M., & Kleinmuntz, D. (1995). Anticipation of effort and accuracy in multi-attribute choice. *Organizational Behavior and Human Decision Processes, 63.1,* 21-32.

Fix, V., Wiedenbeck, S., & Scholtz, J. (1993). Mental representations of programs by novices and experts. *Interchi'93,* 74-79.

Flower, L. (1994). *The Construction of Negotiated. Meaning: A Social Cognitive Theory of Writing.* Carbondale: Southern Illinois University Press.

Foy, P. (1996). The re-invention of the corporate information model. *IEEE Transactions on Professional Communication, 39.1,* 23-29.

Frey, P., & Adesman, P. (1976). Recall memory for visually presented chess positions. *Memory and Cognition, 4,* 541-547.

Gaffney, G. (1999). What is a affinity diagramming? www.infodesign.com.au June 7, 2003.

Ganzach, Y., & Schul, Y. (1995). The influence of quantity of information and goal framing on decisions. *Acta Psychologia, 89*, 23-36.

Ghedira, C., Maret, P., Fayn, J., & Rubel, P. (2002). Adaptive user interface customization through browsing knowledge capitalization. *International Journal of Medical Informatics, 68,* 219-228.

Glick-Smith, J. (2000). The Technical Communicator's Role in the Implementation of Knowledge Management Systems. *STC's 47th Annual Conference*, Orlando, FL, May 21-24, 2000.

Goodwin, S., Huston, K., & Southard, S. (1988). Organization: The key to producing a reader-based software manual. *Proceedings of 35th STC Conference.*

Gribbons, W. (1991). Visual Literacy in Corporate Communication: Some Implications for Information Design. *IEEE Transactions on Professional Communication,* 34.1, 42-50.

Gribbons, W. (1992). Organization by design: Some implications for structuring information. *Journal of Technical Writing and Communication, 22.1,* 57-75.

Gruber, T., Vemuri, S., & Rice, J. (1997). Model-based Virtual document generation. *International Journal of Human-Computer Studies, 46,* 687-706.

Guillemette, R. (1989). Development and validation of a reader-based documentation measure. *International Journal of Man-Machine Studies, 30,* 551-574.

Gutwin, C., & Greenberg, S. (1997). Workspace awareness. Position paper for the ACM CHI'97 Workshop on Awareness in Collaborative Systems. Atlanta, GA, March 22-27.

Hackos, J., & Redish, J. (1998). *User and Task Analysis for Interface Design.* New York: Wiley.

Hackos, J., Hammar, M., & Elser, A. (1997). Customer partnering: Data gathering for complex on-line documentation. *IEEE Transactions on Professional Communication, 40.2*, 102-110.

Hackos, J. (2002). Creating reporting structures through information modeling online: http://www.sapdesignguild.org/community/innovation_articles/edition2/hackos.asp May 17, 2002.

Hakiel, S. (1997). Usability engineering and software engineering: How do they relate. In M. Smith, G. Salvendy & R. Koubek (Eds.), *Proceedings of the 7th International Conference on Human-Computer Interaction* (pp. 521-524). New York: Elsevier.

Hale, D., Sharpe, S., & Haworth, D. (1996). Human-centered. knowledge acquisition: A structural learning theory approach. *International Journal of Human-Computer Studies, 45.4,* 381-396.

Hallgren, C. (1997). Using a problem focus to quickly aid users in trouble. *Proceedings of the 1997 STC Annual Conference.* Washington, DC: STC.

Hancock-Beaulieu, M., & Walker, S. (1992). An evaluation of automatic query expansion in an on-line library catalogue. *Journal of Documentation, 48.4*, 406-421.

Hartley, J., & Trueman, M. (1985). Research strategy for text designers: The role of headings. *Instructional Science, 14,* 99-157.

Hartson, R., & Boehm-Davis, D. (1993). User interface development process and methodologies. *Behavior and Information Technologies, 12.2,* 98-114.

Heath, C., & Luff, P. (1996). Convergent activities: Line control and passenger information on the London underground. In Y. Engeström & D. Middleton (Eds.), *Cognition and Communication at Work (*pp. 96-129). Cambridge UK: Cambridge UP.

Heba, G. (1997). Digital Architectures: A rhetoric of electronic document structures. *IEEE Transactions on Professional Communication, 40.4,* 275-283.

Hefley, W. (1995). Helping users help themselves. *IEEE Software, 12.2*, 93-95.
Henry, P. (1998). *User-Centered Information Design for Improved Software Usability*. Boston: Artech House.
Herbig, P., & Kramer, H. (1992). The phenomenon of innovation overload. *Technology in Society, 14,* 441-461.
Hix, D., & Hartson, R. (1993). *Developing User Interfaces: Ensuring Usability Through Product and Process*. New York: Wiley.
Hodge, C. Chi-Web discussion list. 7/17/2001
Hollnagel, E., & Woods, D. (1983). Cognitive systems engineering: new wine in new bottles. *International Journal of Man-Machine Studies, 18,* 583-600.
Hollnagel, E. (1993). Decision Support and task nets. In G. Klein, J. Orasanu, R. Calderwood & C. Zsambok (Eds.), *Decision Making in Action: Models and Methods* (pp. 31-36) Norwood, NJ: Ablex.
Holtzblatt, K., & Jones, S. (1993). Contextual inquiry: A participatory technique for system design. In D. Schuler & A. Noamioka (Eds.), *Participatory Design: Principles and Practice* (pp. 177-210). Mahwah: Lawrence Erlbaum Associates.
Hook, K. & Svensson, M. (1999). Evaluating adaptive navigation support. In A. Munro, K. Hook & D. Denyon (Eds.), *Social Navigation of Information Space* (pp. 237-249). London: Springer.
Hook, K. (2000). Steps to take before intelligent user interfaces become real. *Interacting with Computers, 12,* 409-426.
Horton, W. (1993). Let's do away with manuals...Before they do away with us. *Technical Communication, 40,* 26-34.
Horton, W. (1990). *Designing and Writing On-line Documentation.* New York: Wiley.
Hsee, C. (1995). Elastic justification: How tempting but task irrelevant factors influence decisions. *Organizational Behavior and Human Decision Processes, 62.3,* 330-337.
Hutchins, E. (1995). *Cognition in the Wild.* Cambridge, MA: MIT Press.
Ingwersen, P. (1994). Polyrepresentation of information needs and sematic entities. *Proceedings of the 17th annual international ACM SIGIR conference on Research and development in information retrieval (pp.* 101-110). Dublin, Ireland.
Interrelations Digraph. (1997). http://www.sytsma.com/tqmtools/relations.htm. April 12, 1997.
Jacobsen, M. J., W. Farrell, et al. (1998). Education in complex systems. Overview of issues presented at the Education in Complex Systems session given at the Second International Conference on Complex Systems, 1998. Available online: http://emergentdesigns.com/mjacobson/ iccs98/ICCS98-REVISED_9-21.pdf
Jirotka, M., & Goguen, J. (1994). Introduction. *Requirements Engineering: Social and Technical Issues.* In M. Jirotka & J. Goguen (Eds.), *Requirements Engineering: Social and Technical Issues* (pp. 1-16). San Diego: Academic Press.
Johannson, R. (1993). *System Modeling and Identification.* Englewood Cliffs, NJ: Prentice-Hall.
Johnson, E., Payne, J., & Bettman, J. (1988). Information displays and preference reversals. *Organizational Behavior and Human Decision Processes, 42,* 1-21.
Johnson, H., & Johnson, P. (1993). Explanation facilities and interactive systems. *Proceedings of Intelligent User Interfaces '93.* Washington, DC: ACM, 159-166.
Johnson, R. (1994). The unfortunate human factor: A selective history of human factors in technical communication. *Technical Communication Quarterly, 3.2*, 195-212.
Johnson, S. (1997). *Interface Culture.* New York: Basic Books.
Johnson-Laird, P. (1983). *Mental Models*. Cambridge UK: Cambridge UP.

Jones, M., Farquhar, J. & Surry, D. (1995). Using metacognitive theories to design user interfaces for computer-based. learning. *Educational Technology, 35.4*, 12-22.

Jones, S. (1999, January 14). Personal conversation.

Kammersgaard, J. (1988). Four different perspective on human-computer interaction. *International Journal of Man-Machine Studies, 28,* 343-362.

Kaplan, C., Fenwick, J., & Chen, J. (1993). Adaptive hypertext navigation based. on user goals and context. *User Modeling and User Adapted. Interaction, 3,* 193-220.

Karat, J., & Bennett, J. (1991). Using scenarios in design meetings—A case study example. In J. Karat (Ed.), *Taking Software Design Seriously: Practical Techniques for Human-Computer Interaction Design* (pp. 63-94). San Diego: Academic Press.

Karat, J. (1995). Scenario use in the design of a speech recognition system. In J. Carroll (Ed.), *Scenario-based. Design: Envisioning Work and Technology in System Development* (pp. 109-134). New York: Wiley.

Kent, T. (1987). Schema theory and technical communication. *Journal of Technical Writing and Communication, 17.3*, 243–252.

Kieras, D., & Polson, P. (1985). An approach to the formal analysis of user complexity. *International Journal of Man-Machine Studies*, *22*. 365-394.

Kieras, D., & Meyer, D. (1998). The role of cognitive task analysis in the application of predictive models of human performance. EPIC Report No. 11 (TR-98/ONR-EPIC-11) March 5, 1998.

King, W., & Teo, T. (1996). Key dimensions of facilitators and inhibitors for the strategic use of information technology. *Journal of Management Information Systems, 12.4,* 35-54.

Klein, G., Orasanu, J., Caldewood, R., & Zsambook, C. (Eds.). (1993). *Decision Making in Action: Model and Methods.* Norwood, NJ: Ablex.

Klein, G. (1988). Do decision biases explain too much. *Human Factors Society Bulletin, 32.5*, 1-3.

Klein, G. (1999). *Sources of Power: How People make Decisions.* Cambridge, MA: MIT.

Klein, G. (1993). A recognition-primed. decision (RPD) model of rapid decision making. In G. Klein, J. Orasanu, R. Calderwood & C. Zsambok (Eds.), *Decision Making in Action: Models and Methods* (pp. 138-147). Norwood, NJ: Ablex.

Kostelnick, C., & Roberts, D. (1998). *Designing Visual Language: Strategies for Professional Communicators*. New York: Allyn and Bacon.

Kumar, H., Plaisant, C., & Shneiderman, B. (1997). Browsing hierarchical data with multi-level dynamic queries and pruning. *International Journal of Human-Computer Studies, 46.1,* 103-124.

Landauer, T. (1995). *The trouble with computers: Usefulness, usability, and productivity.* Cambridge, MA: MIT.

Langholtz, H., Gettys, C., & Foote, B. (1995). Are resource fluctuations anticipated in resource allocation tasks? *Organizational Behavior and Human Decision Processing, 64.3*, 274-282.

Lanning, T. (1991). Let the users design! In J. Karat (Ed.), *Taking Software Design Seriously: Practical Techniques for Human-Computer Interaction Design* (pp. 127-135). San Diego: Academic Press.

Lansdale, M., & Ormerod, T. (1994). *Understanding Interfaces: A Handbook of Human-Computer Dialogue.* London, England: Academic Press.

Laplante, P., & Flaxman, H. (1995). The convergence of technology and creativity in the corporate environment. *IEEE Transactions on Professional Communication, 38.1,* 20-23.

Larichev, O., Olson, D., Moshkovich, H., & Mechitov, A. (1995). Numerical vs cardinal measurements in multi-attribute decision making: How exact is enough? *Organizational Behavior and Human Decision Processing, 64.1,* 9-21.

Larson, K., & Czerwinski, M. (1998). Web page design: Implications of memory, structure, and scent for information retrieval. *Proceedings of CHI '98.* (pp. 25-32). Washington, DC: ACM.

Lay, M. (1989). Nonrhetorical elements of layout and design. In B. Fearing & K. Sparrow, (Eds.), *Technical Writing: Theory and Practice* (pp. 72-85). Modern Language Associates of America, New York.

Lederer, A., & Prasad, J. (1992). Nine management guidelines for better cost estimating. *Communications of the ACM, 35.2,* 51-59.

Lee, K., Leong, H., & Si, A. (1998). Incremental maintenance for dynamic database-derived html pages in digital libraries. *Proceedings of CIKM '98.* Bethesda MD.

Liddle, S., Campbell, D., & Crawford, C. (1999). Automatically extracting structure and data from business reports. *CIKM '99.* (pp. 86-93). Kansas City MO.

Lillies, P. (1991). Some guidelines for making a computer manual more task oriented. *Proceedings of 38th STC Conference.*

Lipson, C. (1988). A social view of technical writing. *Iowa State Journal of Business and Technical Communication, 2.1,* 7-20.

Macaulay, L., Fowler, C., Kirby, M., & Hutt, A. (1990). USTM: A new approach to requirements specification. *Interacting with Computers, 2.1,* 92-118.

Maglio, P., & Matlock, T. (1999). The conceptual structure of information space. In A. Munro, K. Hook, & D. Denyon (Eds.), *Social Navigation of Information Space* (pp. 155 -173). London: Springer

Marchionini, G. (1995). *Information Seeking in Electronic Environments.* New York: Cambridge UP.

Marks, W., & Dulaney, C. (1998). Visual information processing on the World Wide Web. In C. Forsythe, E. Grose, & J. Ratner (Eds.), *Human Factors and Web Development* (pp. 25-43). Mahwah, NJ: Lawrence Erlbaum Associates.

Martin, P. (1986). Computer user documentation problems: Their causes and solutions. *IEEE Transactions on Professional Communications,* PC 29.4.

Mathe, N., & Chen, J. (1996). User-centered indexing for adaptive information access. *User Modeling and User-Adapted. Interaction, 6,* 225-261.

McDermid, J. (1994). Requirements analysis: Orthodoxy, fundamentalism, and heresy. In M. Jirotka & J. Goguen (Eds.), *Requirements Engineering: Social and Technical Issues* (pp. 17-40). San Diego: Academic Press.

McGovern, G. (2003a, October 24). New thinking: Should you centralize or decentralize your web publishing? http://www.gerrymcgovern.com/nt/2003/nt_2003_10_27_publishing.htm.

McGovern, G. (2003b, March 03). New thinking: Why content management software hasn't worked. http://www.gerrymcgovern.com/nt/2003/nt_2003_03_03_cms.htm.

McGovern, G. (2002, September 30). New thinking: Information technology: Trojan Horse of information overload. http://www.gerrymcgovern.com/nt/2002/nt_2002_09_30_trojan.htm.

McGovern, G. (2001, January 29). New thinking: Why THE Web was invented. http://www.gerrymcgovern.com/nt/2001/nt_2001_01_29_why_web_invented.htm.

McGovern, G. (2000, June 4). New thinking: The information virus. http://www.gerrymcgovern.com/nt/2003/2000/nt_2000_06_05_information_virus.htm.

McGovern, G. (2000, December 11). New thinking: In praise of simplicity.

McGovern, G. (1999, July 18). New thinking: Information nobodies. http://www.gerrymcgovern.com/nt/1999/nt_1999_07_19_information_nobodies.htm.

McNamara, D. (2001). Reading both high and low coherence texts: Effects of text sequence and prior knowledge. *Canadian Journal of Experimental Psychology, 55*, 51-62.

McNamara, D., & Kintsch, W. (1996). Learning from text: Effects of prior knowledge and text coherence. *Discourse Processes, 22*, 247-287.

Miller, G. (1956). The magical number seven, plus or minus two: Some limits on our capacity for processing information. *The Psychological Review, 63*, 81-97.

Miller-Jacobs, H. (1991). Rapid prototyping: An effective technique for system development. In J. Karat (Ed.), *Taking Software Design Seriously: Practical Techniques for Human-Computer Interaction Design* (pp. 273-286). San Diego: Academic Press.

Mirel, B. (1988). Cognitive processing, text linguistics and documentation writing. *Journal of Technical Writing and Communication*, *18.2*, 111-133.

Mirel, B. (2003). Dynamic usability: Designing usefulness into systems for complex tasks. In M. Albers & B. Mazur (Eds.), *Content and Complexity: Information Design in Software Development and Documentation* (pp. 233-261). Mahwah, NJ: Lawrence Erlbaum Associates.

Mirel, B. (1993). Beyond the monkey house: Audience analyses in computerized. workplaces. In R. Spilka (Ed.), *Writing in the Workplace: New Research Perspectives* (pp. 21-40). Carbondale: Southern Illinois UP.

Mirel, B. (1992). Analyzing audiences for software manuals: A survey of instructional needs for 'real world tasks.' *Technical Communication Quarterly, 1.1*, 15-35.

Mirel, B. (1998). Applied constructivism for user documentation. *Journal of Business and Technical Communication, 12.1,* 7-49.

Mirel, B. (1996). Writing and database technology: Extending the definition of writing in the workplace. In P. Sullivan and J. Dautermann. (Eds.), *Electronic Literacies in the Workplace.* (pp. 91-114). Urbana, IL: NCTE.

Morville, P. (2001). Software for information architects. http://argus-acia.com/strange_connections/strange011.html

Mumby, D. (1988). *Communication and Power in Organizations: Discourse, Ideology, and Domination*. Norwood, NJ: Ablex.

Murray, D. (1995). *Knowledge Machines: Language and Information in a Technological Society.* New York: Longman.

Myers, B. (1994, January). Challenges of HCI design and implementation. *Interactions,* 73-83.

Nanard, J., & Nanard, M. (1995) Hypertext design environments and the hypertext design process. *Communications of the ACM, 38.8,* 49-56.

Nardi, B., Miller, J., & Wright, D. (1998). Collaborative, programmable intelligent agents. *Communications of the ACM, 41.3,* 96-104.

Neerincx, M. & Griffioen, E. (1996). Cognitive task analysis: Harmonizing tasks to human capacities. *Ergonomics*, *39.4,* 543-561.

Nielson, J. Alertbox: Current issues in web usability. http://www.useit.com/alertbox/

Nelson, T. Project Xanadu . http://www.xanadu.net/

Norman, D. (1986). Cognitive engineering. In D. Norman & S. Draper (Eds.), *User Centered System Design: New Perspectives on Human-computer Interaction* (pp. 33-62). Mahwah, NJ: Lawrence Erlbaum Associates.

Norman, D. (1984). Stages and levels in human-machine interaction. *International Journal of Man-Machine Studies, 21*, 365-375.

Norman, D. (1990). *The Design of Everyday Things.* Currency.

Norman, D. (1998). *The Invisible Computer.* Cambridge, MA: MIT Press.

Norman, D. (1993). *Things that make us smart: Defending human attributes in the age of the machine.* Reading, MA: Addison-Wesley.

Odell, L. (1985). Beyond the text: Relations between writing and social context. In L. Odell & D. Goswami (Eds.), *Writing in Nonacademic Settings* (pp. 249-280). New York: Guilford.

O'Malley, C. (1986). Helping users help themselves. In D. Norman & S. Draper (Eds.), *User Centered System Design: New Perspectives on Human-computer Interaction* (pp. 377-398). Mahwah, NJ: Lawrence Erlbaum Associates.

Orasanu, J., & Connolly, T. (1993). The reinvention of decision making. In G. Klein, J. Orasanu, R. Calderwood & C. Zsambok (Eds.), *Decision Making in Action: Models and Methods* (pp. 3-20). Norwood, NJ: Ablex.

Orlikowski, W. & Gash, D. (1994). Technological frames: Making sense of information technology in organizations. *ACM Transactions on Information Systems, 12.2,* 174-207.

Outsell Inc. (2000). Study reveals the business-to-business content market as a hidden industry. http://www.outsellinc.com/cgi-bin/devsite/press.pl?sizings1 June 2, 2000.

Owen, D. (1986). Answers first, then questions. In D. Norman & S. Draper (Eds.), *User Centered System Design: New Perspectives on Human-computer Interaction* (pp. 362-375). Mahwah, NJ: Lawrence Erlbaum Associates.

Paradis, J., Dobrin, D., & Miller, R. (1985). Writing at Exxon ITD: Notes on the writing environment of an R&D organization. In L. Odell & D. Goswami (Eds.), *Writing in Nonacademic Settings* (pp. 281-308). New York: Guilford.

Parker, H., Roast, C., & Siddiqi, J. (1997). HCI and requirements engineering: Towards a framework for investigating temporal properties in interaction. *SIGCHI Bulletin,* 29.1 http://www.acm.org/sigchi/bulletin/1997.1 May 4, 1997.

Ponte, J., & Croft, B. (1998). A language modeling approach to information retrieval. *SIGIR '93*, Melbourne, Australia.

Quesenbery, W. (2003). Dimensions of usability. In M. Albers & B. Mazur (Eds.), *Content and Complexity: Information Design in Software Development and Documentation* (pp. 81-102). Mahwah, NJ: Lawrence Erlbaum Associates.

Rainer, K., & Watson, H. (1996). The keys to executive information system success. *Journal of Management Information Systems, 12.2,* 83-98.

Randel, J., & Pugh, L. (1996). Differences in expert and novice situation awareness in naturalistic decision making. *International Journal of Human-Computer Studies, 45,* 579-597.

Rasmussen, J., Pejtersen, A., & Goodstein, L. (1994). *Cognitive Systems Engineering.* New York: Wiley.

Rasmussen, J. (1986). *Information Processing and Human-Machine Interaction: An Approach to Cognitive Engineering.* New York: North-Holland.

Ray, W., Hess, S., & Goldberg, J. (1997). Mapping physiology to cognition in adaptive interface design. In M. Smith, G. Salvendy & R. Koubek (Eds.), *Proceedings of the 7th International Conference on Human-Computer Interaction* (pp. 491-494). New York: Elsevier.

Redish, J. (1985). Making information accessible. In L. Odell & D. Goswami (Eds.), *Writing in Nonacademic Settings* (pp. 129-153). New York: Guilford.

Redish, J. & Schell, D. (1989). Writing and testing instructions for usability. In B. Fearing & K. Sparrow (Eds.), *Technical Writing: Theory and Practice* (pp. 63-71). Modern Language Associates of America, New York.

Redish, J. (1994). Understanding readers. In C. Barnum & S. Carliner (Eds.), *Techniques For Technical Communicators* (pp. 15-41). New York: Macmillan.

Redish, J. (1998). Minimalism in technical communication: some issues to consider. In John Carroll. (Ed.). *Minimalism beyond the Nurnberg Funnel*. Cambridge, MA: MIT.

Rein, G., McCue, D., & Slein, J. (1997). A case for document management functions on the web. *Communications of the ACM, 40.9,* 81-89.

Roberts, D. (1989). Readers' comprehension responses in informative discourse: Toward connecting reading and writing in technical communication. *Journal of Technical Writing and Communication, 19.2,* 135-148.

Robertson, G., Card, S., & Mackinlay, J. (1993). Information visualization: Using 3D interactive animation. *Communications of the ACM, 36.4,* 57-71.

Rockley Group. (2002). Enterprise content management through single sourcing. www.rockley.com.

Rockley, A. (2003). Information modeling. In M. Albers & B. Mazur (Eds.), *Content and Complexity: Information Design in Software Development and Documentation* (pp. 331-360). Mahwah, NJ: Lawrence Erlbaum Associates.

Rockley, A., Kostur, P., & Manning, S. (2002). *Managing Enterprise Content: A Unified Content Strategy* Indianapolis, IN: Que.

Rosenbaum, S. & Walters, D. (1986). Audience diversity: A major challenge in computer documentation. *IEEE Transactions on Professional Communications, PC-29.4*, 48-55.

Rosenfeld, L., & Morville, P. (1998). *Information Architecture for the World Wide Web.* Cambridge, MA: O'Reilly.

Roth, E. M., Bennett, K. B., & Woods, D. D. (1987). Human interaction with an intelligent machine. *International Journal of Man-Machine Studies*, *27*, 479-526.

Rouse, W., & Valusek, J. (1993). Evolutionary design of systems to support decision making. In G. Klein, J. Orasanu, R. Calderwood & C. Zsambok (Eds.), *Decision Making in Action: Models and Methods* (pp. 270-286). Norwood, NJ: Ablex.

Rubens, P., & Rubens, B. (1988). Usability and format design. In S. Doheny-Farina (Ed.). *Effective Documentation: What We Have Learned. From Research* (pp. 213-234). Cambridge, MA: MIT.

Ruskin, J. (2000). *The Humane Interface: New Directions for Designing Interactive Systems.* New York: Addison-Wesley.

Santhanam, R., & Wiedenbeck, S. (1993). Neither novice nor expert: The discretionary user of software. *International Journal of Man-Machine Studies, 38,* 201-229.

Santoni, C., Furtado, E., & Francois, P. (1997). Aid methodology for designing adaptive human computer interfaces for supervision systems. In M. Smith, G. Salvendy, & R. Koubek (Eds.), *Proceedings of the 7th International Conference on Human-Computer Interaction* (pp. 501-504). New York: Elsevier.

Schamber, L. (1995). A user-based cognitive approach to modeling highly dynamic information problem situations. *Proceedings of the 58th Annual Meeting of the American Society for Information Science, 32*, 157-162. http://www.asis.org/asis-95/papers/schamber.html

Schriver, K. (1996). *Dynamics in Document Design.* New York: Wiley.

Shao, H. (2001). The evolution of content management. In G. McGovern (ed.). *Content Critical* http://www.gerrymcgovern.com/cc_forum.htm. November 8, 2001.

Shneiderman, B. (1992). *Designing the User Interface: Strategies for Effective Human-Computer Interaction.* Reading, MA: Addison-Wesley.

Simon, H. (1979). *Models of Thought.* New Haven, CT: Yale UP.

Slovic, P. (1972, April). From Shakespeare to Simon: Speculations—and some evidence—about man's ability to process information. *Oregon Research Institute Research Bulletin.*

Smart, G. (1993). Genre as community invention. In R. Spilka (Ed.), *Writing in the Workplace: New Research Perspectives* (pp. 124-134). Carbondale: Southern Illinois UP.

Smith, E., & Goodman, L. (1984). Understanding written instructions: The role of an explanatory schema. *Cognition and Instruction, 1.4,* 359-396.

So, R., Finney, C., Tseng, M., Su, C., & Yen, B. (1997). A closed-loop approach for integrating human factors into system development. In M. Smith, G. Salvendy & R. Koubek (Eds.), *Proceedings of the 7th International Conference on Human-Computer Interaction* (pp. 517-520). New York: Elsevier.

Spolsky, J. (2003). User interface design for programmers: Chapter 2: Figuring out what they expected. http://www.joelonsoftware.com/uibook/chapters/fog0000000058.html. March 25, 2003.

Spool, J. (2003). 5 things to know about users. http://www.uiconf.com/7west/five_things_to_know_article.htm June 17, 2003.

Sprague, R. (1995). Electronic document management: Challenges and opportunities for information system managers. *MIS Quarterly, 19.1,* 29-50.

Spyridakis, J., & Wenger, M. (1992). Writing for human performance: Relating reading research to document design. *Technical Communication, 39.2*, 202-215.

Steinberg, L., & Gitomer, D. (1993). Cognitive task analysis, interface design, and technical troubleshooting. *Proceedings of Intelligent User Interfaces '93.* Washington, DC: ACM, 185-190.

Suchman, L. (1987). *Plans and Situated Actions: The Problem of Human-Machine Communication.* Cambridge, UK: Cambridge UP.

Sullivan, K., & Kida, T. (1995). The effect of multiple reference points and prior gains and losses on manager's risky decision making. *Organizational Behavior and Human Decision Processing, 64.1,* 76-83.

Sumner, T., Bonnardel, N., & Kallak, B. (1997). The cognitive ergonomics of knowledge-based design support systems. *Proceedings of CHI'97.* Washington, DC: ACM, 83-90.

Sutcliffe, A. (1997). Task-related. information analysis. *International Journal of Human-Computer Studies, 47*, 223-257.

Takahaski, M., Takei, S., & Kitamura, M. (1997). Multimodal display for enhanced situation awareness. In M. Smith, G. Salvendy & R. Koubek (Eds.), *Proceedings of the 7th International Conference on Human-Computer Interaction* (pp. 707-710). New York: Elsevier.

Tannenbaum, A. (2002). *Metadata solutions: Using metamodels, repositories, XML and enterprise portals to generate information on demand.* Boston, MA: Addison-Wesley.

Tebeaux, E. (1996). Nonacademic writing into the 21st century: Achieving and sustaining relevance in research and curricula. In. A. Duin & C. Hansen (Eds.), *Nonacademic Writing: Social Theory and Technology* (pp. 35-55) Mahwah, N J: Lawrence Erlbaum Associates.

Terveen, L., Selfridge, P., & Long, D. (1995). Living design memory: Framework, implementation, lessons learned. *Human-Computer Interaction, 10.1*, 1-37.

Terwilliger, R., & Polson, P. (1997). Relationships between users' and interfaces' task representations. *Proceedings of CHI'97.* Washington, DC: ACM, 99-106.

Thompson, H. & Coney, M. (1995). Putting reader roles to the test: An ethnomethodological approach. *IEEE Transactions on Professional Communication, 38.2*, 100-109.

Thuring, M., Hannemann, J., & Haake, J. (1995). Hypermedia and cognition: Designing for comprehension. *Communications of the ACM, 38*, 57-66.

Treu, S., Mullins, P., & Adams, J. (1989). A network-wide information system: Multi-level context for the user at the workstation interface. *Information Systems, 14.5,* 393-406.

Treu, S. (1990). Conceptual distance and interface-supported: Visualization of information objects and patterns. *Journal of Visual Languages and Computing, 1*, 369-388.

Treu, S. (1992). Interface structures: Conceptual, logical, and physical patterns applicable to human-computer interaction. *International Journal of Man-Machine Studies, 37,* 565-593.

Tufte, E. (1983). *The Visual Display of Quantitative Information.* Cheshire, CT: Graphics Press.

Tversky, A., & Kahneman, D. (1981). The framing of decisions an the psychology of choice. *Science, 211,* 453-458.

User Interface Engineering. (1998). Techniques for complex applications http://world.std.com/~uieweb/index.html April 24, 1998.

van der Meij, H., Blijleven, P., & Jansen, L. (2003). What makes up a procedure? In M. Albers & B. Mazur (Eds.), *Content and Complexity: Information Design in Software Development and Documentation* (pp. 129-186). Mahwah, NJ: Lawrence Erlbaum Associates.

Vicente, K. (1999a). *Cognitive Work Analysis.* Mahway, New Jersey: Lawrence Erlbaum Associates.

Vicente, K. (1999b). Wanted: Psychologically relevant, device- and event-independent work analysis techniques. *Interacting with Computers, 11,* 237-254.

Visser, W., & Morals, A. (1990). Concurrent use of different expertise elicitation methods applied to the study of the programming activity. In D. Ackermann & M. Tauber (Eds.), *Mental Models and Human-Computer Interactions* (pp. 97-114). Amsterdam: North Holland.

Waern, Y. (1989). *Cognitive Aspects of Computer Supported. Tasks.* New York: Wiley.

Warren, T. (1993). Three approaches to reader analysis. *Technical Communication, 40.1*, 81-87.

Webster, D., & Kruglanski, A. (1994). Individual differences in need. for cognitive closure. *Journal of Personality and Social Psychology, 67.6*, 1049 -1672.

Weiss, E. (2002). Egoless writing: Improving quality by replacing artistic impulse with engineering discipline. *Journal of Computer Documentation, 26.1*, 3-10.

Weiss, E. (1993). Of document databases, SGML, and rhetorical neutrality. *IEEE Transactions on Professional Communication, 36.2,* 58-61.

Whitburn, M. (1984). The ideal orator and literary critic as technical communicators: An emerging revolution in english departments. In A. Lundsford & L. Ede (Eds.), *Essays in Classical Rhetoric and Modern Discourse* (pp. 230-248). Carbondale, IL: Southern Illinois University Press.

Wickens, C., & Carswell, C. M. (1995). The proximity compatibility principle: Its psychological foundation and relevance to display design. *Human Factors, 37.3,* 473-494.

Wickens, C. (1992). *Engineering Psychology and Human Performance.* New York: HarperCollins.

Wilson, E. (2003). Asynchronous health care communication. *Communications of the ACM, 46.6,* 79-84.

Woods, D., & Roth. E. (1988). Cognitive engineering: Human problem solving with tools. *Human Factors, 30.4,* 415-430.

Wright, P. (1974). The harassed decision maker: Time pressure, distractions, and the use of evidence. *Journal of Applied. Psychology, 59.5,* 555-561.

Yazici, H., Muthuswamy, K., & Vila, J. (1994). An intelligent system approach for graphical user interface management. *Journal of Educational Multimedia and Hypermedia, 3.1,* 37-54.

Yoon, W., & Hammer, J. (1988a). Aiding operator during novel fault diagnosis. *IEEE Transactions on Systems, Man, and Cybernetics, 18.1,* 142-147.

Yoon, W., & Hammer, J. (1988b). Deep-reasoning fault diagnosis: An aid and a model. *IEEE Transactions on Systems, Man, and Cybernetics, 18.4*, 659-675.

# Author Index

# I



# J



# K



# L



# M

# N



# O



# P



# Q



# R

## S



## T



## U



## V



## W

## Y



## Z

# Subject Index

## P



## Q



## R



## S